A Practical Introduction to Optical Mineralogy

TITLES OF RELATED INTEREST

Boninites
A. J. Crawford (ed.)

Carbonatites
K. Bell (ed.)

Cathodoluminescence of geological materials
D. J. Marshall

Chemical fundamentals of geology
R. Gill

Crystal structures and cation sites of the rock-forming minerals
J. R. Smyth & D. L. Bish

The dark side of the Earth
R. Muir Wood

Deformation processes in minerals, ceramics and rocks
D. J. Barber & H. Maurer

High-temperature metamorphism and crustal anatexis
J. Ashworth & M. Brown (eds)

Igneous petrogenesis
B. M. Wilson

Image interpretation in geology
S. Drury

The inaccessible Earth
G. C. Brown & A. E. Mussett

The interpretation of igneous rocks
K. G. Cox et al.

Introduction to X-ray spectrometry
K. L. Williams

Komatites
N. Arndt & E. Nisbet (eds)

Mathematics in geology
J. Ferguson

Petrology of the igneous rocks
F. Hatch et al.

Petrology of the metamorphic rocks
R. Mason

Rheology of the Earth
G. Ranalli

Rutley's elements of mineralogy
C. D. Gribble

Stimulating the Earth
J. Holloway & B. Wood

Statistical methods in geology
R. F. Cheeney

Volcanic successions
R. Cas & J. V. Wright

The young Earth
E. G. Nisbet

A Practical Introduction to Optical Mineralogy

C. D. Gribble
Department of Geology, University of Glasgow

A. J. Hall
Department of Applied Geology, University of Strathclyde

CHAPMAN & HALL
London · Glasgow · New York · Tokyo · Melbourne · Madras

Published by Chapman & Hall, 2-6 Boundary Row,
London SE1 8HN, UK

Chapman & Hall, 2-6 Boundary Row, London SE1 8HN, UK

Blackie Academic & Professional, Wester Cleddens Road, Bishopbriggs, Glasgow G64 2NZ, UK

Chapman & Hall Inc., One Penn Plaza, 41st Floor, New York, NY 10119, USA

Chapman & Hall Japan, Thomson Publishing Japan, Hirakawacho Nemoto Building, 6F, 1-7-11 Hirakawa-cho, Chiyoda-ku, Tokyo 102, Japan

Chapman & Hall Australia, Thomas Nelson Australia, 102 Dodds Street, South Melbourne, Victoria 3205, Australia

Chapman & Hall India, R. Seshadri, 32 Second Main Road, CIT East, Madras 600 035, India

First edition 1985
Reprinted 1991, 1992, 1994

© 1985 C.D. Gribble and A.J. Hall

Typeset in 9/11 Times by D.P. Media Limited, Hitchin, Herts
Printed in Great Britain by St Edmundsbury Press Ltd, Bury St Edmunds, Suffolk

ISBN 0 412 53280 8

Apart from any fair dealing for the purposes of research or private study, or criticism or review, as permitted under the UK Copyright Designs and Patents Act, 1988, this publication may not be reproduced, stored, or transmitted, in any form or by any means, without the prior permission in writing of the publishers, or in the case of reprographic reproduction only in accordance with the terms of the licences issued by the Copyright Licensing Agency in the UK, or in accordance with the terms of licences issued by the appropriate Reproduction Rights Organization outside the UK. Enquiries concerning reproduction outside the terms stated here should be sent to the publishers at the London address printed on this page.
    The publisher makes no representation, express or implied, with regard to the accuracy of the information contained in this book and cannot accept any legal responsibility or liability for any errors or omissions that may be made.

A catalogue record for this book is available from the British Library
Library of Congress Cataloging-in-Publication Data available

Preface

Microscopy is a servant of all the sciences, and the microscopic examination of minerals is an important technique which should be mastered by all students of geology early in their careers. Advanced modern textbooks on both optics and mineralogy are available, and our intention is not that this new textbook should replace these but that it should serve as an introductory text or a first stepping-stone to the study of optical mineralogy. The present text has been written with full awareness that it will probably be used as a laboratory handbook, serving as a quick reference to the properties of minerals, but nevertheless care has been taken to present a systematic explanation of the use of the microscope as well as theoretical aspects of optical mineralogy. The book is therefore suitable for the novice either studying as an individual or participating in classwork.

Both transmitted-light microscopy and reflected-light microscopy are dealt with, the former involving examination of transparent minerals in thin section and the latter involving examination of opaque minerals in polished section. Reflected-light microscopy is increasing in importance in undergraduate courses on ore mineralisation, but the main reason for combining the two aspects of microscopy is that it is no longer acceptable to neglect opaque minerals in the systematic petrographic study of rocks. Dual purpose microscopes incorporating transmitted- and reflected-light modes are readily available, and these are ideal for the study of polished thin sections. The technique of preparing polished thin sections has been perfected for use in the electron microprobe analyser, which permits analysis of points of the order of one micron diameter on the polished surface of the section. Reflected-light study of polished thin sections is a prerequisite of electron microprobe analysis, so an ability to characterise minerals in reflected light is of obvious advantage. Reflected-light microscopy is described with consideration of the possibility that experienced transmitted-light microscopists may wish to use this book as an introduction to the reflected-light technique.

This book therefore introduces students to the use of both the transmitted- and reflected-light microscope and to the study of minerals using both methods (Ch. 1). The descriptive section on minerals is subdivided for ease of presentation: the silicates (which are studied using transmitted light) are described in Chapter 2, and are followed in Chapter 3 by the non-silicates (which are studied using either transmitted or reflected light). The minerals are presented in alphabetical order but, to save duplicating descriptions, closely related minerals have been presented together. The best way to locate the description of a given mineral is therefore to look up the required mineral in the index, where minerals appear in alphabetical order. Although important, a detailed understanding of optical theory is not essential to mineral identification.

ACKNOWLEDGEMENTS

Accounts of transmitted-light optical crystallography and reflected-light theory are therefore placed after the main descriptions of minerals, in Chapters 4 and 5 respectively. The appendices include systematic lists of the optical properties of minerals for use in identification.

This book is intended to be an aid to the identification of minerals under the microscope, but not to the description or interpretation of mineral relationships. We both hope that the text fills its intended slot, and that students find it helpful and enjoyable to use.

Acknowledgements

The sections dealing with transmitted light have been written by C. D. Gribble. He acknowledges the debt owed to Kerr (1977), whose format has generally been employed in Chapter 2, and to Deer et al. (1966), whose sections on physical properties and mineral paragenesis have often been the basis of the RI values and occurrences given in this text. Other authors and papers have been employed, in particular Smith (1974) on the feldspars and Wahlstrom (1959) on optical crystallography.

Descriptions of the opaque minerals by A. J. Hall are based on data in many texts. However, they are taken mainly from the tables of Uytenbogaardt and Burke (1971), the classic text *Dana's system of mineralogy* edited by Palache et al. (1962), the unsurpassed description of the textures of the ore minerals by Ramdohr (1969), and the atlas by Picot and Johan (1977). The textbook on the microscopic study of minerals by Galopin and Henry (1972), and course notes and publications of Cervelle, form the basis of the section on theoretical aspects of reflected-light microscopy.

The Michel–Levy chart on the back cover is reproduced with the kind permission of Carl Zeiss of Oberkochen, Federal Republic of Germany.

We are grateful for support and suggestions by our colleagues in the Universities of Glasgow and Strathclyde. A special thanks is due to the typists Janette Forbes, Irene Wells, Dorothy Rae, Irene Elder and Mary Fortune.

Also, we are particularly grateful to John Wadsworth and Fergus Gibb for their comments and reviews of the original manuscript, and to Brian Goodale for his comments on Chapter 4.

Any errors or inaccuracies are, however, ours.

Contents

Preface *page*	vii
Acknowledgements	viii
List of tables	xi
List of symbols and abbreviations used in text	xii

1 Introduction to the microscopic study of minerals

1.1	Introduction	1
1.2	The transmitted-light microscope	1
1.3	Systematic description of minerals in thin section using transmitted light	5
	1.3.1 Properties in plane polarised light	5
	1.3.2 Properties under crossed polars	8
1.4	The reflected-light microscope	12
1.5	The appearance of polished sections under the reflected-light microscope	17
1.6	Systematic description of minerals in polished section using reflected light	19
	1.6.1 Properties observed using plane polarised light (PPL)	19
	1.6.2 Properties observed using crossed polars	20
	1.6.3 The external nature of grains	21
	1.6.4 Internal properties of grains	21
	1.6.5 Vickers hardness number (VHN)	22
	1.6.6 Distinguishing features	22
1.7	Observations using oil immersion in reflected-light studies	22
1.8	Polishing hardness	23
1.9	Microhardness (VHN)	25
1.10	Points on the use of the microscope (transmitted and reflected light)	26
1.11	Thin- and polished-section preparation	28

2 Silicate minerals

2.1	Crystal chemistry of silicate minerals	30
2.2	Mineral descriptions	34
	Al_2SiO_5 polymorphs 35; Amphibole group 41; Beryl 56; Chlorite 57; Chloritoid 58; Clay minerals 59; Cordierite 61; Epidote group 63; Feldspar group 67; Feldspathoid family 84; Garnet group 87; Humite group 88; Mica group 90; Olivine group 95; Pumpellyite 98; Pyroxene group 99;	

Scapolite 117; Serpentine 119; Silica group 120; Sphene 124; Staurolite 125; Talc 126; Topaz 127; Tourmaline group 128; Vesuvianite 129; Zeolite group 129; Zircon 131

3 The non-silicates

3.1	Introduction	page 132
3.2	Carbonates	132
3.3	Sulphides	138
3.4	Oxides	154
3.5	Halides	167
3.6	Hydroxides	169
3.7	Sulphates	171
3.8	Phosphate	175
3.9	Tungstate	175
3.10	Arsenide	176
3.11	Native elements	177

4 Transmitted-light crystallography

4.1	Polarised light: an introduction	180
4.2	Refractive index	181
4.3	Isotropy	181
4.4	The biaxial indicatrix triaxial ellipsoid	183
4.5	The uniaxial indicatrix	184
4.6	Interference colours and Newton's Scale	186
4.7	Fast and slow components, and order determination	190
	4.7.1 Fast and slow components	190
	4.7.2 Quartz wedge and first order red accessory plate	191
	4.7.3 Determination of order of colour	191
	4.7.4 Abnormal or anomalous interference colours	192
4.8	Interference figures	192
	4.8.1 Biaxial minerals	192
	4.8.2 Sign determination for biaxial minerals	196
	4.8.3 Flash figures	196
	4.8.4 Uniaxial minerals	197
	4.8.5 Isotropic minerals	197
4.9	Pleochroic scheme	197
	4.9.1 Uniaxial minerals	197
	4.9.2 Biaxial minerals	200
4.10	Extinction angle	200

5 Reflected-light theory

5.1	Introduction	page	202
	5.1.1 Reflectance		203
	5.1.2 Indicating surfaces of reflectance		206
	5.1.3 Observing the effects of crystallographic orientation on reflectance		206
	5.1.4 Identification of minerals using reflectance measurements		209
5.2	Colour of minerals in PPL		209
	5.2.1 CIE (1931) colour diagram		210
	5.2.2 Exercise on quantitative colour values		211
5.3	Isotropic and anisotropic sections		212
	5.3.1 Isotropic sections		212
	5.3.2 Anisotropic sections		213
	5.3.3 Polarisation colours		213
	5.3.4 Exercise on rotation after reflection		215
	5.3.5 Detailed observation of anisotropy		216

Appendix A.1	Refractive indices of biaxial minerals	218
Appendix A.2	Refractive indices of uniaxial positive minerals	219
Appendix A.3	Refractive indices of uniaxial negative minerals	220
Appendix A.4	Refractive indices of isotropic minerals	221
Appendix B	$2V$ and sign of biaxial minerals	222
Appendix C	Properties of ore minerals	225
Appendix D	Mineral identification chart	237
Appendix E	Gangue minerals	239
Bibliography		241
Index		243

List of tables

1.1	Optical data for air and oil immersion	page	22
1.2	Relation between VHN and Moh's hardness		27
3.1	Optical properties of the common carbonates		134
3.2	Spinels		158
4.1	Extinction angle sections not coincident with maximum birefringence sections		201

List of symbols and abbreviations used in the text

Crystallographic properties of minerals

$a\,b\,c$ or $X\,Y\,Z$	crystallographic axes
hkl	Miller's indices, which refer to crystallographic orientation
(111)	a single plane or face
{111}	a form; all planes with same geometric relationship to axes
[111]	zone axis; planes parallel to axis belong to zone
β	angle between a and c in the monoclinic system
α, β, γ	angles between b and c, a and c, and a and b in the triclinic system

Light

λ	wavelength
A	amplitude
PPL	plane or linearly polarised light

Microscopy

N, S, E, W	north (up), south (down), east (right), west (left) in image or in relation to crosswires
NA	numerical aperture
XPOLS, XP, CP	crossed polars (analyser inserted)

Optical properties

n or RI	refractive index of mineral
N	refractive index of immersion medium
n_o	RI of ordinary ray
n_e	RI of extraordinary ray
n_α	minor RI
n_β	intermediate
n_γ	major RI
o	ordinary ray vibration direction of uniaxial mineral
e	extraordinary ray vibration direction of uniaxial mineral
α, β, γ	principal vibration directions of general optical indicatrix
δ	maximum birefringence ($n_e - n_o$ or $n_\gamma - n_\alpha$)
$2V$	optic axial angle
$2V_\alpha$	optic axial angle bisected by α
$2V_\gamma$	optic axial angle bisected by γ
Bx_a	acute bisectrix (an acute optic axial angle)
Bx_o	obtuse bisectrix (an obtuse optic axial angle)
OAP	optic axial plane
$\gamma\hat{\,}cl$	angle between γ (slow component) and cleavage
$\alpha\hat{\,}cl$	angle between α (fast component) and cleavage
k	absorption coefficient
R	reflectance (usually expressed as a percentage, R%)
R_{min}	minimum reflectance of a polished section of a bireflecting mineral grain
R_{max}	maximum reflectance of a polished section of a bireflecting mineral grain

SYMBOLS AND ABBREVIATIONS

R_o	principal reflectance corresponding to ordinary ray vibration direction of a uniaxial mineral
R_e	principal reflectance corresponding to extraordinary ray vibration direction of a uniaxial mineral
ΔR	bireflectance ($R_{max} - R_{min}$) referring to individual section or maximum for mineral

Quantitative colour

$Y\%$	visual brightness
λ_d	dominant wavelength
$P_e\%$	saturation
x, y	chromaticity co-ordinates

Mineral properties

VHN	Vickers hardness number
H	hardness on Moh's scale
D	density
SG	specific gravity

General

P	pressure
T	temperature
XRD	X-ray diffraction
REE	rare earth elements
nm	nanometre
μm	micrometre or micron
mm	millimetre
cm	centimetre
d	distance or length
Å	angstroms
cl	cleavage
kb	kilobar
>	greater than
<	less than
$\geq$	greater than or equal to
$\leq$	less than or equal to
$\sim$	approximately
$\approx$	approximately equal to
$\perp$	perpendicular to
$\parallel$	parallel to
4+	four or greater
3D	three dimensional
Zn + Fe + S	association of elements in ternary chemical system
Zn − Fe − S	association of elements

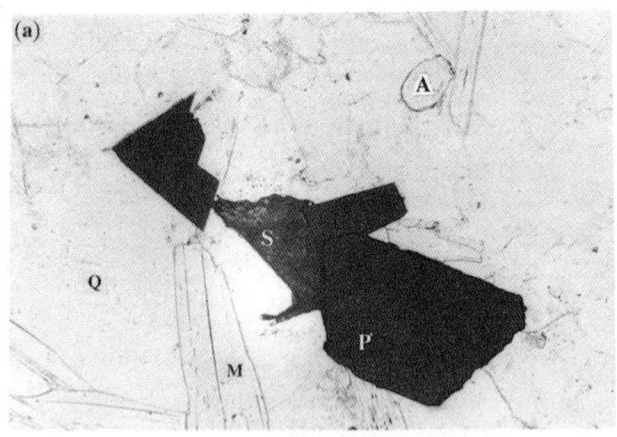

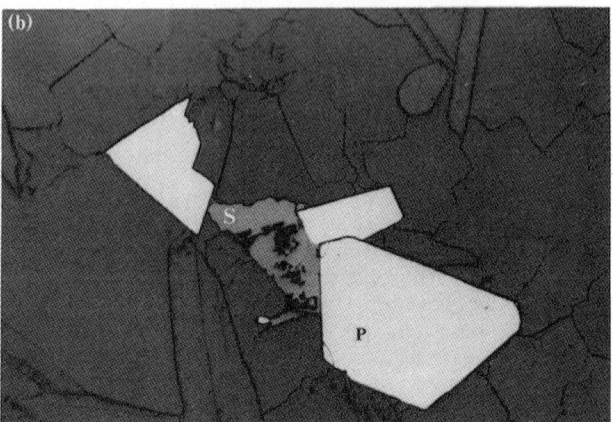

Frontispiece Photomicrographs, taken using (a) transmitted light and (b) reflected light, of the same area of a polished thin section of quartzite containing pyrite (P), sphalerite (S), muscovite (M), apatite (A) and abundant quartz (Q).

The features illustrated in transmitted light are: (i) opacity – pyrite is the only opaque phase, sphalerite is semi-opaque, and the others are transparent; (ii) relief – very high (sphalerite, $n \approx 2.4$), moderate (apatite, $n \approx 1.65$), moderate (muscovite, $n \approx 1.60$), and low (quartz, $n \approx 1.55$); (iii) cleavage – perfect in muscovite (n is the refractive index of the mineral).

The feature illustrated in reflected light is reflectance: 54% (pyrite, white – true colour slightly yellowish white), 17% (sphalerite, grey), 6% (apatite, dark grey), 5% (muscovite, dark grey), and 5% (quartz, dark grey) (reflectance is the percentage of incident light reflected by the mineral).

Note that opaque grains, grain boundaries and cleavage traces appear black in transmitted light, whereas pits (holes), grain boundaries and cleavage traces appear black in reflected light.

1 Introduction to the microscopic study of minerals

1.1 Introduction

Microscopes vary in their design, not only in their appearance but also in the positioning and operation of the various essential components. These components are present in all microscopes and are described briefly below. Although dual purpose microscopes incorporating both transmitted- and reflected-light options are now available (Fig. 1.1), it is more convenient to describe the two techniques separately. More details on the design and nature of the components can be obtained in textbooks on microscope optics.

1.2 The transmitted-light microscope

The light source
In transmitted-light studies a lamp is commonly built into the microscope base (Fig. 1.2). The typical bulb used has a tungsten filament (A source) which gives the field of view a yellowish tint. A blue filter can be inserted immediately above the light source to change the light colour to that of daylight (C source).

In older microscopes the light source is quite separate from the microscope and is usually contained in a hooded metal box to which can be added a blue glass screen for daylight coloured light. A small movable circular mirror, one side of which is flat and the other concave, is attached to the base of the microscope barrel. The mirror is used to direct the light through the rock thin section on the microscope stage, and the flat side of the mirror should be used when a condenser is present.

The polariser
The assumption is that light consists of electromagnetic vibrations. These vibrations move outwards in every direction from a point source of 'white' light, such as a microscope light. A polarising film (the polariser) is held within a lens system located below the stage of the microscope, and this is usually inserted into the optical path. On passing through the polariser the light is 'polarised' and now vibrates in a single

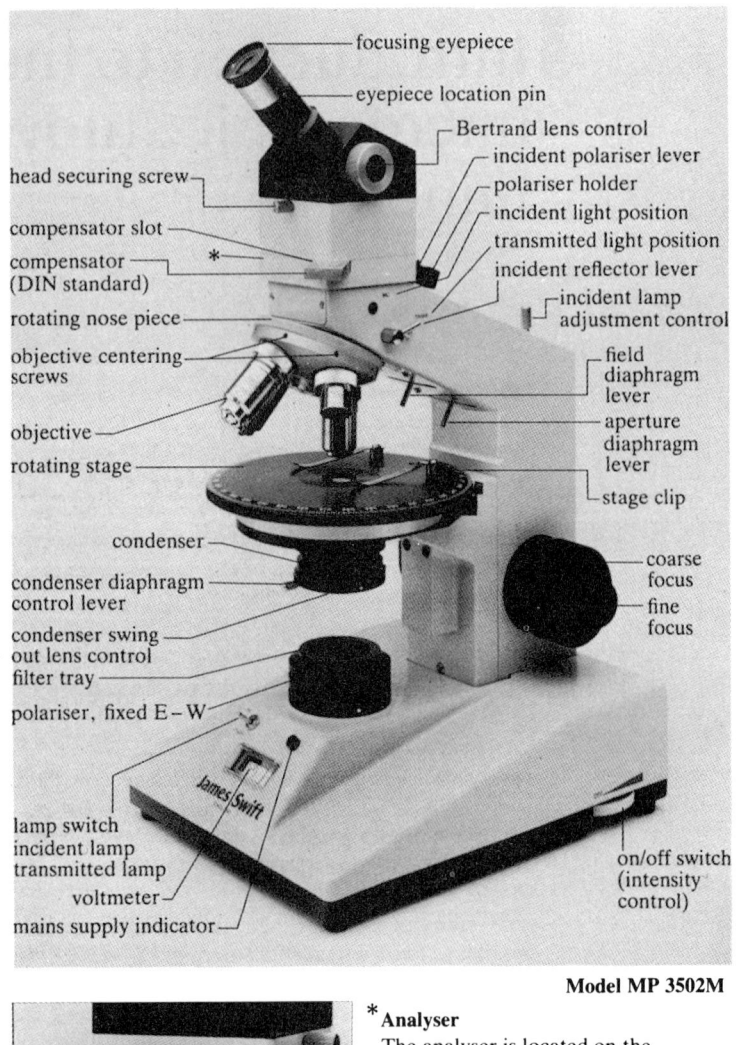

Figure 1.1 The Swift Student polarising microscope (photo courtesy of Swift Ltd).

THE TRANSMITTED-LIGHT MICROSCOPE

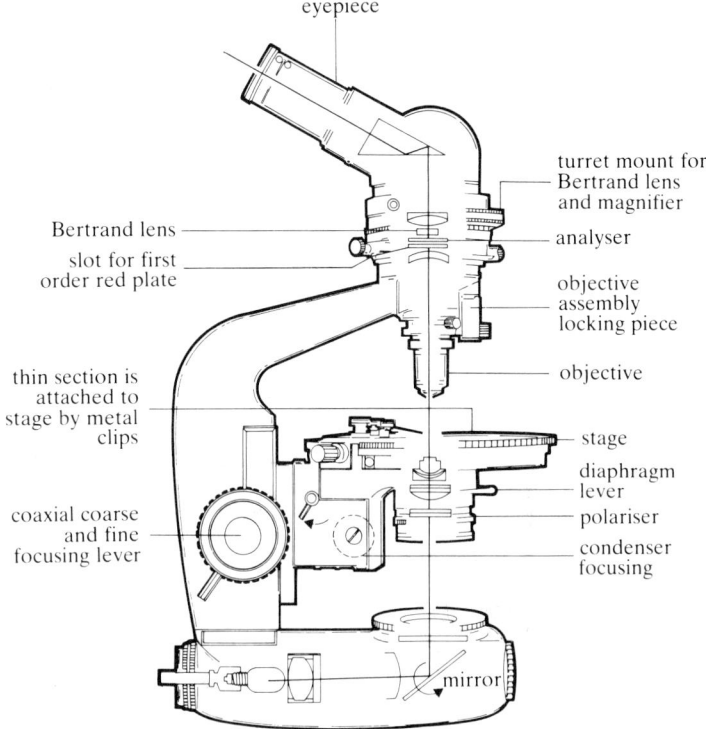

Figure 1.2 Modern transmitted light microscope. Older models may focus by moving the upper barrel of the microscope (not the stage as in the illustration), and may use an external light source. The illustration is based on a Nikon model POH-2 polarising microscope.

plane. This is called *plane polarised light* (PPL). In most UK microscopes the polariser is oriented to give E–W vibrating incident light (see also Ch. 4).

Substage diaphragms
One or two diaphragms may be located below the stage. The field diaphragm, often omitted on simple student microscopes, is used to reduce the area of light entering the thin section, and should be in focus at the same position as the thin section; it should be opened until it just disappears from view. The aperture diaphragm is closed to increase resolution; it can be seen when the Bertrand lens is inserted.

The condenser or convergent lens
A small circular lens (the condenser) is attached to a swivel bar, so that it can be inserted into the optical train when required. It serves to direct a cone of light on to the thin section and give optimum resolution for the

objectives used. The entire lens system below the microscope stage, including polariser, aperture diaphragm and condenser, can often be racked upwards or downwards in order to optimise the quality of illumination. Some microscopes, however, do not possess a separate convergent lens and, when a convergent lens is needed, the substage lens system is racked upwards until it is just below the surface of the microscope stage.

Stage
The microscope stage is flat and can be rotated. It is marked in degrees, and a side vernier enables angles of rotation to be accurately measured. The stage can usually be locked in place at any position. The rock thin section is attached to the centre of the stage by metal spring clips.

Objectives
Objectives are magnifying lenses with the power of magnification inscribed on each lens (e.g. ×5, ×30). An objective of very high power (e.g. ×100) usually requires an immersion oil between the objective lens and the thin section.

Eyepiece
The eyepiece (or ocular) contains crosswires which can be independently focused by rotating its uppermost lens. Eyepieces of different magnification are available. Monocular heads are standard on student microscopes. Binocular heads may be used and, if correctly adjusted, reduce eye fatigue.

The analyser
The analyser is similar to the polariser; it is also made of polarising film but oriented in a N–S direction, i.e. at right angles to the polariser. When the analyser is inserted into the optical train, it receives light vibrating in an E–W direction from the polariser and cannot transmit this; thus the field of view is dark and the microscope is said to have *crossed polars* (CP, XPOLS or XP). With the analyser out, the polariser only is in position; plane polarised light is being used and the field of view appears bright.

The Bertrand lens
This lens is used to examine interference figures (see Section 1.3.2). When it is inserted into the upper microscope tube an interference figure can be produced which fills the field of view, provided that the convergent lens is also inserted into the optical path train.

The accessory slot
Below the analyser is a slot into which accessory plates, e.g. quartz wedge, or first order red, can be inserted. The slot is oriented so that

SYSTEMATIC DESCRIPTION OF MINERALS

accessory plates are inserted at 45° to the crosswires. In some microscopes the slot may be rotatable.

Focusing

The microscope is focused either by moving the microscope stage up or down (newer models) or by moving the upper microscope tube up or down (older models). Both coarse and fine adjusting knobs are present.

1.3 Systematic description of minerals in thin section using transmitted light

Descriptions of transparent minerals are given in a particular manner in Chapters 2 and 3, and the terms used are explained below. The optical properties of each mineral include some which are determined in plane polarised light, and others which are determined with crossed polars. For most properties a low power objective is used (up to $\times 10$).

1.3.1 Properties in plane polarised light

The analyser is taken out of the optical path to give a bright image (see Frontispiece).

Colour

Minerals show a wide range of colour (by which we mean the natural or 'body' colour of a mineral), from colourless minerals (such as quartz and feldspars) to coloured minerals (brown biotite, yellow staurolite and green hornblende). Colour is related to the wavelength of visible light, which ranges from violet (wavelength $\lambda = 0.00039$ mm or 390 nm) to red ($\lambda = 760$ nm). White light consists of all the wavelengths between these two extremes. With colourless minerals in thin section (e.g. quartz) white light passes unaffected through the mineral and none of its wavelengths is absorbed, whereas with opaque minerals (such as metallic ores) all wavelengths are absorbed and the mineral appears black. With coloured minerals, selective absorption of wavelengths take place and the colour seen represents a combination of wavelengths of light transmitted by the mineral.

Pleochroism

Some coloured minerals change colour between two 'extremes' when the microscope stage is rotated. The two extremes in colour are each seen twice during a complete (360°) rotation. Such a mineral is said to be pleochroic, and ferromagnesian minerals such as the amphiboles, biotite and staurolite of the common rock-forming silicates possess this property.

Pleochroism is due to the unequal absorption of light by the mineral in different orientations. For example, in a longitudinal section of biotite, when plane polarised light from the polariser enters the mineral which has its cleavages parallel to the vibration direction of the light, considerable absorption of light occurs and the biotite appears dark brown. If the mineral section is then rotated through 90° so that the plane polarised light from the polariser enters the mineral with its cleavages now at right angles to the vibration direction, much less absorption of light occurs and the biotite appears pale yellow.

Habit
This refers to the shape that a particular mineral exhibits in different rock types. A mineral may appear euhedral, with well defined crystal faces, or anhedral, where the crystal has no crystal faces present, such as when it crystallises into gaps left between crystals formed earlier. Other descriptive terms include prismatic, when the crystal is elongate in one direction, or acicular, when the crystal is needle like, or fibrous, when the crystals resemble fibres. Flat, thin crystals are termed tabular or platy.

Cleavage
Most minerals can be cleaved along certain specific crystallographic directions which are related to planes of weakness in the mineral's atomic structure. These planes or cleavages which are straight, parallel and evenly spaced in the mineral are denoted by Miller's indices, which indicate their crystallographic orientation. Some minerals such as quartz and garnet possess no cleavages, whereas others may have one, two, three or four cleavages. When a cleavage is poorly developed it is called a parting. Partings are usually straight and parallel but *not* evenly spaced. The number of cleavages seen depends upon the orientation of the mineral section. Thus, for example, a prismatic mineral with a square cross section may have two prismatic cleavages. These cleavages are seen to intersect in a mineral section cut at right angles to the prism zone, but in a section cut parallel to the prism zone the traces of the two cleavages are parallel to each other and the mineral appears to possess only one cleavage (e.g. pyroxenes, andalusite).

Relief
All rock thin sections are trapped between two thin layers of resin (or cementing material) to which the glass slide and the cover slip are attached. The refractive index (RI) of the resin is 1.54. The surface relief of a mineral is essentially constant (except for carbonate minerals), and depends on the difference between the RI of the mineral and the RI of the enclosing resin. The greater the difference between the RI of the mineral and the resin, the rougher the appearance of the surface of the mineral. This is because the surfaces of the mineral in thin section are

made up of tiny elevations and depressions which reflect and refract the light.

If the RIs of the mineral and resin are similar the surface appears smooth. Thus, for example, the surfaces of garnet and olivine which have much higher RIs than the resin appear rough whereas the surface of quartz, which has the same RI as the resin (1.54) is smooth and virtually impossible to detect.

To obtain a more accurate estimate of the RI of a mineral (compared to 1.54) a mineral grain should be found at the edge of the thin section, where its edge is against the cement. The diaphragm of the microscope should be adjusted until the edge of the mineral is clearly defined by a thin, bright band of light which is exactly parallel to the mineral boundary. The microscope tube is then carefully racked upwards (or the stage lowered), and this thin band of light – the Becke line – will appear to move towards the medium with the higher RI. For example, if $RI_{mineral}$ is greater than Ri_{cement} the Becke line will appear to move into the mineral when the microscope tube is slowly racked upwards. If the RI of a mineral is close to that of the cement then the mineral surface will appear smooth and dispersion of the refractive index may result in slightly coloured Becke lines appearing in both media. The greater the difference between a mineral's RI and that of the enclosing cement, the rougher the surface of the mineral appears. An arbitrary scheme used in the section of mineral descriptions is as follows:

RI	Description of relief
1.40–1.50	moderate
1.50–1.58	low
1.58–1.67	moderate
1.67–1.76	high
>1.76	very high

The refractive indices of adjacent minerals in the thin section may be compared using the Becke line as explained.

Alteration

The most common cause of alteration is by water or CO_2 coming into contact with a mineral, chemically reacting with some of its elements, and producing a new, stable mineral phase(s). For example, water reacts with the feldspars and produces clay minerals. In thin section this alteration appears as an area of cloudiness within the transparent feldspar grain. The alteration may be so advanced that the mineral is completely replaced by a new mineral phase. For example, crystals of olivine may have completely altered to serpentine, but the area occupied by the serpentine still has the configuration of the original olivine crystal. The olivine is said to be pseudomorphed by serpentine.

THE MICROSCOPIC STUDY OF MINERALS

1.3.2 Properties under crossed polars

The analyser is inserted into the optical path to give a dark, colourful image.

Isotropism

Minerals belonging to the cubic system are isotropic and remain dark under crossed polars whatever their optical orientation. All other minerals are anisotropic and usually appear coloured and go into extinction (that is, go dark) four times during a complete rotation of the mineral section. This property, however, varies with crystallographic orientation, and each mineral possesses at least one orientation which will make the crystal appear to be isotropic. For example, in tetragonal, trigonal and hexagonal minerals, sections cut perpendicular to the c axis are always isotropic.

Birefringence and interference colour

The colour of most anisotropic minerals under crossed polars varies, the same mineral showing different colours depending on its crystallographic orientation. Thus quartz may vary from grey to white, and olivine may show a whole range of colours from grey to red or blue or green. These are colours on Newton's Scale, which is divided into several orders:

Order	Colours
first	grey, white, yellow, red
second	violet, blue, green, yellow, orange, red
third	indigo, green, blue, yellow, red, violet
fourth and above	pale pinks and green

A Newton's Scale of colours can be found on the back cover of this book. These orders represent interference colours; they depend on the thickness of the thin section mineral and the birefringence, which is the difference between the two refractive indices of the anisotropic mineral grain. The thin section thickness is constant (normally 30 microns) and so interference colours depend on birefringence; the greater the birefringence, the higher the order of the interference colours. Since the maximum and minimum refractive indices of any mineral are oriented along precise crystallographic directions, the highest interference colours will be shown by a mineral section which has both maximum and minimum RIs in the plane of the section. This section will have the maximum birefringence (denoted δ) of the mineral. Any differently oriented section will have a smaller birefringence and show lower colours. The descriptive terms used in Chapter 2 are as follows:

SYSTEMATIC DESCRIPTION OF MINERALS

Maximum birefringence (δ)	Interference colour range	Description
0.00–0.018	first order	low
0.018–0.036	second order	moderate
0.036–0.055	third order	high
> 0.055	fourth order or higher	very high

Very low may be used if the birefringence is close to zero and the mineral shows anomalous blue colours.

Interference figures

Interference figures are shown by all minerals except cubic minerals. There are two main types of interference figures (see Figs 4.19 and 21), uniaxial and biaxial.

Uniaxial figures may be produced by suitably orientated sections from tetragonal, trigonal and hexagonal minerals. An isotropic section (or near isotropic section) of a mineral is first selected under crossed polars, and then a high power objective ($\times 40$ or more) is used with the substage convergent lens in position and the aperture diaphragm open. When the Bertrand lens is inserted into the optical train a black cross will appear in the field of view. If the cross is off centre, the lens is rotated so that the centre of the cross occurs in the SW (lower left hand) segment of the field of view.

The first order red accessory plate is then inserted into the optical train in such a way that the length slow direction marked on it points towards the centre of the black cross, and the colour in the NE quadrant of the cross is noted:

blue means that the mineral is positive   (denoted +ve)
yellow means that the mineral is negative   (denoted −ve)

Some accessory plates are length fast, and the microscope may not allow more than one position of insertion. In this case the length fast direction will point towards the centre of the black cross and the colours and signs given above would be reversed, with a yellow colour meaning that the mineral is positive and a blue colour negative. It is therefore essential to appreciate whether the accessory plate is length fast or slow, and how the fast or slow directions of the accessory plate relate to the interference figure after insertion (see Fig. 4.20).

Biaxial figures may be produced by suitable sections of orthorhombic, monoclinic and triclinic minerals. An isotropic section of the mineral under examination is selected and the microscope mode is as outlined for uniaxial figures, i.e. $\times 40$ objective and convergent lens in position. Inserting the Bertrand lens will usually reveal a single optic axis interference figure which appears as a black arcuate line (or isogyre) crossing the field of view. Sometimes a series of coloured ovals will appear, arranged about a point on the isogyre, especially if the mineral section is very thick or if the mineral birefringence is very high. The stage is then rotated until the isogyre is in the 45° position (relative to the crosswires) and concave towards the NE segment of the field of view. In this position the isogyre curvature can indicate the size of the optic axial angle ($2V$) of a mineral. The more curved the isogyre the smaller the $2V$. The curvature will vary from almost a 90° angle, indicating a very low $2V$ (less than 10°) to 180° when the isogyre is straight (with a $2V$ of 80° to 90°). When the $2V$ is very small (less than 10°) both isogyres will be seen in the field of view, and the interference figure resembles a uniaxial cross, which breaks up (i.e. the isogyres move apart) on rotation. The first order red accessory plate (length slow) is inserted and the colour noted on the *concave* side of the isogyre:

> blue means that the mineral is positive ($+$ve)
> yellow means that the mineral is negative ($-$ve)

If the accessory plate is length fast (as mentioned in the preceding section) the colours above will be reversed, that is a yellow colour will be positive and blue negative (see Fig. 4.20).

Extinction angle
Anisotropic minerals go into extinction four times during a complete 360° rotation of a mineral section. If the analyser is removed from the optical train while the mineral grain is in extinction, the orientation of some physical property of the mineral, such as a cleavage or trace of a crystal face edge, can be related to the microscope crosswires.

All uniaxial minerals possess *straight* or *parallel* extinction since a prism face or edge, or a prismatic cleavage, or a basal cleavage, is parallel to one of the crosswires when the mineral is in extinction.

Biaxial minerals possess either *straight* or *oblique* extinction. Orthorhombic minerals (olivine, sillimanite, andalusite, orthopyroxenes) show straight extinction against either a prismatic cleavage or a prism face edge. All other biaxial minerals possess oblique extinction, although in some minerals the angular displacement may be extremely small: for example, an elongate section of biotite showing a basal cleavage goes into extinction when these cleavages are almost parallel to one of the microscope crosswires. The angle through which the mineral has then to be rotated to bring the cleavages parallel to the crosswire will vary from nearly 0° to 9° depending on the biotite composition, and this angle is called the *extinction angle*.

The maximum extinction angle of many biaxial minerals is an important optical property and has to be precisely determined. This is done as follows. A mineral grain is rotated into extinction, and the angular position of the microscope stage is noted. The polars are uncrossed (by removing the upper analyser from the optical train) and the mineral grain rotated until a cleavage trace or crystal trace edge or twin plane is parallel to the crosswires in the field of view. The position of the microscope stage is again noted and the difference between this reading and the former one gives the extinction angle of the mineral grain. Several grains are tested since the crystallographic orientation may vary and the *maximum extinction angle* obtained is noted for that mineral. The results of measurements from several grains should *not* be averaged.

Extinction angles are usually given in mineral descriptions as the angle between the slow (γ) or fast (α) ray and the cleavage or face edge (written as γ or $\alpha\hat{\,}$cl), and this technique is explained in detail in Chapter 4.

In many biaxial minerals the maximum extinction angle is obtained from a mineral grain which shows maximum birefringence such as, for example, the clinopyroxenes diopside, augite and aegirine, and the monoclinic amphiboles tremolite and the common hornblendes. However, in some minerals the maximum extinction angle is not found in a section showing maximum birefringence. This is so for the clinopyroxene pigeonite, the monoclinic amphiboles crossite, katophorite and arfvedsonite, and a few other minerals of which kyanite is the most important (see also Ch. 4, Section 4.10).

Throughout the mineral descriptions given in Chapter 2, large variations in the maximum extinction angle are shown for particular minerals. For example the maximum extinction angles for the amphiboles tremolite–actinolite are given as between 18° and 11° ($\gamma\hat{\,}$cleavage). Tremolite, the Mg-rich member, has a maximum extinction angle between 21° and 17°, whereas ferroactinolite has a maximum extinction angle from 17° to 11°. This variation in the extinction angle is caused mainly by variations in the Mg:Fe ratio. Variation in extinction angles are common in many minerals or mineral pairs which show similar chemical changes.

Twinning
This property is present when areas with differing extinction orientations within the same mineral grain have planar contacts. Often only a single twin plane is seen, but in some minerals (particularly plagioclase feldspars) multiple or lamellar twinning occurs with parallel twin planes.

Zoning
Compositional variation (zoning) within a single mineral may be expressed in terms of changes of 'natural' colour from one zone to an adjoining one; or by changes in birefringence; or by changes in extinction orientation. These changes may be abrupt or gradational, and commonly occur as a sequence from the core of a mineral grain (the early-formed part) to its edge (the last-formed part).

Zoning is generally a growth phenomenon and is therefore related to the crystal shape.

Dispersion
Refractive index increases as the wavelength of light decreases. Thus the refractive index of a mineral for red light is less than for blue light (since the wavelength of red light is greater than the wavelength of blue light). White light entering a mineral section is split into the colours of the spectrum, with blue nearest to the normal (i.e. the straight through path) and red the furthest away. This breaking up of the white light is called *dispersion*. In most minerals the amount of dispersion is very small and will not affect the mineral's optical properties. However, the Na-rich clinopyroxenes, the Na-rich amphiboles, sphene, chloritoid, zircon and brookite possess very strong dispersion. With many of these minerals, interference figures may be difficult to obtain and the use of accessory plates (to determine mineral sign etc.) may not be possible.

Each mineral possesses a few diagnostic properties, and in the descriptions in Chapter 2 these have been marked with an asterisk. Sometimes a final paragraph discusses differences between the mineral being described and other minerals that have similar optical properties.

1.4 The reflected-light microscope

The light source
A high intensity source (Fig. 1.3) is required for reflected-light studies, mainly because of the low brightness of crossed polar images. Tungsten-halogen quartz lamps are used, similar to those in transparency projectors, and the tungsten light (A source) gives the field a yellowish tint. Many microscopists prefer to use a blue correction filter to change the light colour to that of daylight (C source). A monochromatic light source (coloured light corresponding to a very limited range of the visible spectrum) is rarely used in qualitative microscopy, but monochromatic filters for the four standard wavelengths (470 nm, 546 nm, 589 nm and 650 nm) could be useful in comparing the brightness of coexisting minerals, especially now that quantitative measurements of brightness are readily available.

The polariser
Polarised light is usually obtained by using a polarising film, and this should be protected from the heat of the lamp by a glass heat filter. The polariser should always be inserted in the optical train. It is best fixed in orientation to give E–W vibrating incident light. However, it is useful to be able to rotate the polariser on occasion in order to correct its orientation or as an alternative to rotating the analyser.

THE REFLECTED-LIGHT MICROSCOPE

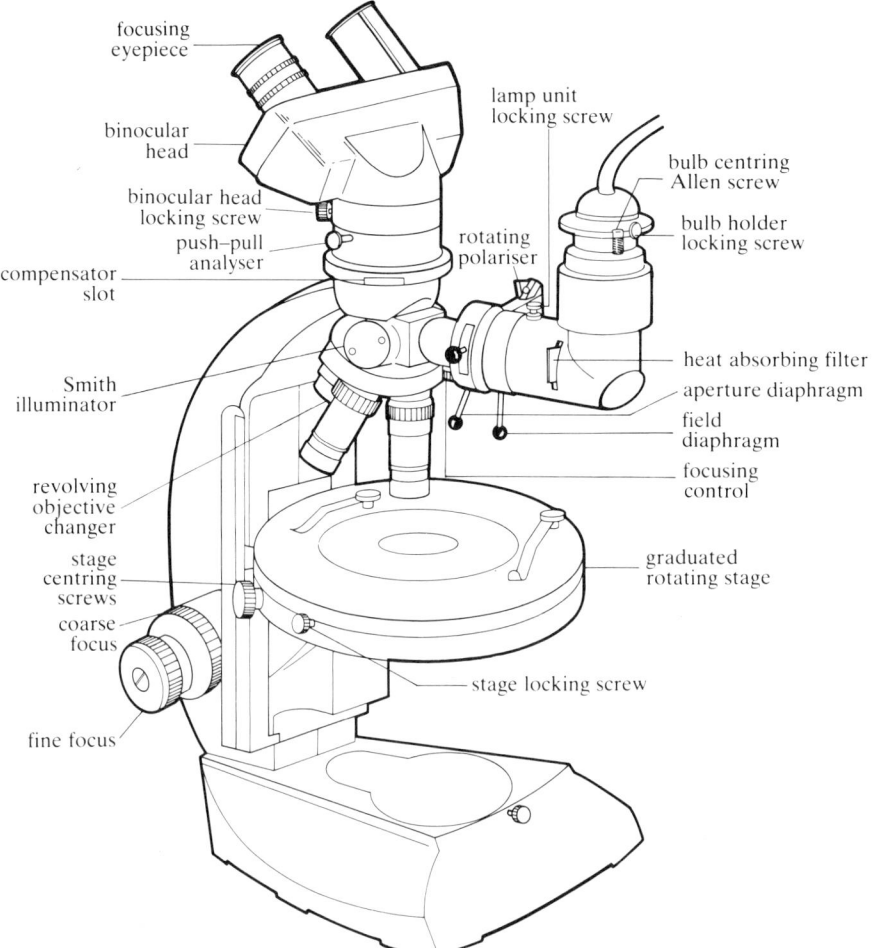

Figure 1.3 The Vickers M73 reflected light microscope. Note that it is the *polariser* that rotates in this microscope.

The incident illuminator

The incident illuminator sits above the objective and its purpose is to reflect light down through the objective on to the polished specimen. As the reflected light travels back up through the objective to the eyepiece it must be possible for this light to pass through the incident illuminator. There are three types of reflector used in incident illuminators (Fig. 1.4):

(a) The cover glass or coated thin glass plate (Fig. 1.4a). This is a simple device, but is relatively inefficient because of light loss both before and after reflection from the specimen. However, its main disadvantage when at 45° inclination is the lack of uniform extinction of an isotropic field. This is due to rotation of the vibration direction of polarised reflected light which passes asymmetrically through the cover glass on returning towards the eyepiece. This disadvantage is overcome by decreasing the angle to about 23° as on Swift microscopes.

(b) The mirror plus glass plate or Smith illuminator (Fig. 1.4b). This is slightly less efficient than the cover glass but, because of the low angle (approaching perpendicular) of incidence of the returning reflected light on the thin glass plate, extinction is uniform and polarisation colours are quite bright. This illuminator is used on Vickers microscopes.

(c) The prism or total reflector (Fig. 1.4c). This is more efficient than the glass plate type of reflector but it is expensive. It would be 100 per cent efficient, but half of the light flux is lost because only half of the aperture of the objective is used. A disadvantage is the lack of uniform extinction obtained. A special type of prism is the triple prism or Berek prism, with which very uniform extinction is obtained because of the nature of the prism (Hallimond 1970, p. 103). Prism reflectors are usually only available on research microscopes and are normally interchangeable with glass plate reflectors. One of the disadvantages of the prisms is that the incident light is slightly oblique, and this can cause a shadow effect on surfaces with high relief. Colouring of the shadow may also occur.

Figure 1.4 Incident illuminators.

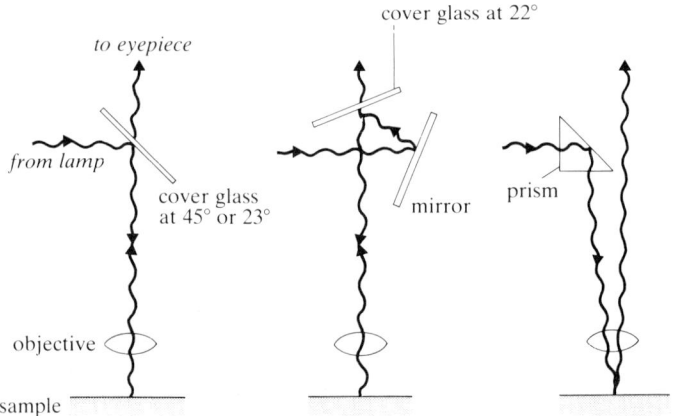

(a) **Cover glass illuminator**   (b) **Smith illuminator**   (c) **Prism illuminator**

THE REFLECTED-LIGHT MICROSCOPE

Objectives

Objectives are magnifiers and are therefore described in terms of their magnification power, e.g. ×5. They are also described using numerical aperture (Fig. 1.5), the general rule being the higher the numerical aperture the larger the possible magnification. It is useful to remember that, for objectives described as being of the same magnification, a higher numerical aperture leads to finer resolved detail, a smaller depth of focus and a brighter image. Objectives are designed for use with *either* air (dry) *or* immersion oil between the objective lens and the sample. The use of immersion oil between the objective and sample leads to an increase in the numerical aperture value (Fig. 1.5). Immersion objectives are usually engraved as such.

Low power objectives can usually be used for either transmitted or reflected light, but at high magnifications (> ×10) good images can only be obtained with the appropriate type of objective. Reflected-light objectives are also known as metallurgical objectives. Achromatic objectives are corrected for chromatic aberration, which causes colour fringes in the image due to dispersion effects. Planochromats are also corrected for spherical aberration, which causes a loss in focus away from the centre of a lens; apochromats are similarly corrected but suffer from chromatic difference of magnification, which must be removed by use of compensating eyepieces.

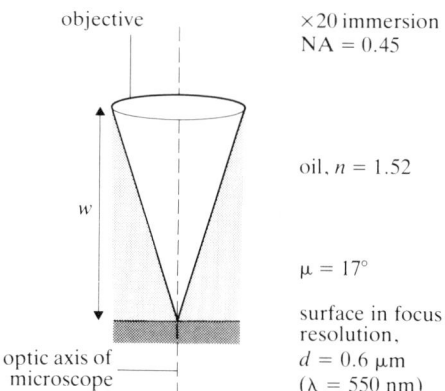

Figure 1.5 Numerical aperture and resolution. N.A. = $n \sin \mu$, where N.A. = numerical aperture, n = refractive index of immersion medium, and μ = half the angle of the light cone entering the objective lens (for air, $n = 1.0$). $d = 0.5 \lambda/\text{N.A.}$ where d = the resolution (the distance between two points that can be resolved) and λ is in microns (1 micron = 1000 nm). The working distance (w in the diagram) depends on the construction of the lens; for the same magnification, oil immersion lenses usually have a shorter distance than dry objectives.

Analyser
The analyser may be moved in and out of the optical train and rotated through small angles during observation of the specimen. The reason for rotation of the analyser is to enhance the effects of anisotropy. It is taken out to give plane polarised light (PPL), the field appearing bright, and put in to give crossed polars (XPOLS), the field appearing dark. Like the polariser, it is usually made of polarising film. On some microscopes the analyser is fixed in orientation and the polariser is designed to rotate. The effect is the same in both cases, but it is easier to explain the behaviour of light assuming a rotating analyser (Section 5.3).

The Bertrand lens
This is usually little used in reflected-light microscopy, especially by beginners. The polarisation figures obtained are similar, but differ in origin and use, to the interference figures of transmitted-light microscopy.

Isotropic minerals give a black cross which is unaffected by rotation of the stage but splits into two isogyres on rotation of the analyser. Lower symmetry minerals give a black cross in the extinction position, but the cross separates into isogyres on rotation of either the stage or the analyser. Colour fringes on the isogyres relate to dispersion of the rotation properties.

Light control
Reflected-light microscopes are usually designed to give Kohler-type critical illumination (Galopin & Henry 1972, p. 58). As far as the user is concerned, this means that the aperture diaphragm and the lamp filament can be seen using conoscopic light (Bertrand lens in) and the field diaphragm can be seen using orthoscopic light (Bertrand lens out).

A lamp rheostat is usually available on a reflected-light microscope to enable the light intensity to be varied. A very intense light source is necessary for satisfactory observation using crossed polars. However, for PPL observations the rheostat is best left at the manufacturer's recommended value, which should result in a colour temperature of the A source. The problem with using a decreased lamp intensity to decrease image brightness is that this changes the overall colour of the image. Ideally, neutral density filters should be used to decrease brightness if the observer finds it uncomfortable. In this respect, binocular microscopes prove less wearisome on the eyes than monocular microscopes.

Opening of the *aperture diaphragm* decreases resolution, decreases the depth of focus and increases brightness. It should ideally be kept only partially open for PPL observation but opened fully when using crossed polars. If the aperture diaphragm can be adjusted, it is viewed using the Bertrand lens or by removing the ocular (eyepiece). Figure 1.6

Figure 1.6
Centring of the aperture diaphragm.

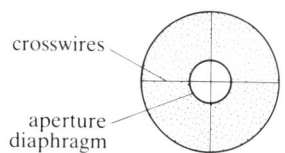

Correctly centred aperture diaphragm for a plate glass reflector
image with Bertrand lens inserted and aperture diaphragm closed

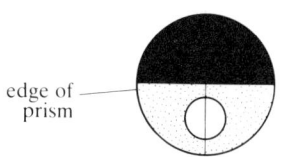

Correctly centred aperture diaphragm for a prism reflector
image with Bertrand lens inserted and aperture diaphragm closed

shows the aperture diaphragm correctly centred for glass plate and prism reflectors.

The *illuminator field diaphragm* is used simply to control scattered light. It can usually be focused and should be in focus at the same position as the specimen image. The field diaphragm should be opened until it just disappears from the field of view.

1.5 The appearance of polished sections under the reflected-light microscope

On first seeing a polished section of a rock or ore sample the observer often finds that interpretation of the image is rather difficult. One reason for this is that most students use transmitted light for several years before being introduced to reflected light, and they are conditioned into interpreting bright areas as being transparent and dark areas as being opaque; for polished sections the opposite is the case! It is best to begin examination of a polished section such as that illustrated in Figure 1.7 by using *low power magnification* and *plane polarised light*, when most of the following features can be observed:

(a) Transparent phases appear dark grey. This is because they reflect only a small proportion of the incident light, typically 3 to 15%. Occasionally bright patches are seen within areas of transparent minerals, and are due to reflection from surfaces under the polished surface.

(b) Absorbing phases (opaques or ore minerals) appear grey to bright white as they reflect much more of the incident light, typically 15 to 95%. Some absorbing minerals appear coloured, but usually colour tints are very slight.

THE MICROSCOPIC STUDY OF MINERALS

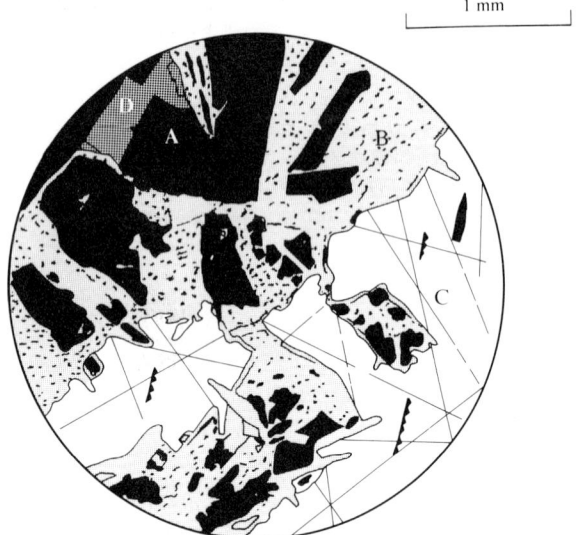

Figure 1.7 Diagrammatic representation of a polished section of a sample of lead ore. *Transparent phases*, e.g. fluorite (A), barite (B) and the mounting resin (D) appear dark grey. Their brightness depends on their refractive index. The fluorite is almost black. *Absorbing phases (opaque)*, e.g. galena (C), appear white. *Holes, pits and cracks* appear black. Note the black triangular cleavage pits in the galena and the abundant pits in the barite which results, not from poor polishing, but from the abundant fluid inclusions. *Scratches* appear as long straight or curving lines. They are quite abundant in the galena which is soft and scratches easily.

(c) Holes, pits, cracks and specks of dust appear black. Reflection from crystal faces in holes may give peculiar effects such as very bright patches of light.

(d) Scratches on the polished surface of minerals appear as long straight or curving lines, often terminating at grain boundaries or pits. Severe fine scratching can cause a change in the appearance of minerals. Scratches on native metals, for example, tend to scatter light and cause colour effects.

(e) Patches of moisture or oil tend to cause circular dark or iridescent patches and indicate a need for cleaning of the polished surface.

(f) Tarnishing of minerals is indicated by an increase in colour intensity, which tends to be rather variable. Sulphides, for example bornite, tend to tarnish rapidly. Removal of tarnishing usually requires a few minutes buffing or repolishing.

(g) Polishing relief, due to the differing hardnesses of adjacent minerals, causes dark or light lines along grain contacts. Small soft bright grains may appear to glow, and holes may have indistinct dark margins because of polishing relief.

SYSTEMATIC DESCRIPTION OF MINERALS

1.6 Systematic description of minerals in polished section using reflected light

Most of the ore minerals described in Chapter 3 have a heading 'polished section'. The properties presented under this heading are in a particular sequence, and the terms used are explained briefly below. Not all properties are shown by each mineral, so only properties which might be observed are given in Chapter 3.

1.6.1 Properties observed using plane polarised light (PPL)

The analyser is taken out of the optical path to give a bright image (see Frontispiece).

Colour

Most minerals are only slightly coloured when observed using PPL, and the colour sensation depends on factors such as the type of microscope, the light source and the sensitivity of an individual's eyes. Colour is therefore usually described simply as being a variety of grey or white, e.g. bluish grey rutile, pinkish white cobaltite.

Pleochroism

If the colour of a mineral varies from grain to grain and individual grains change in colour on rotation of the stage, then the mineral is pleochroic. The colours for different crystallographic orientations are given when available. Covellite, for example, shows two extreme colours, blue and bluish light grey. Pleochroism can often be observed only by careful examination of groups of grains in different crystallographic orientation. Alternatively the pleochroic mineral may be examined adjacent to a non-pleochroic mineral, e.g. ilmenite against magnetite.

Reflectance

This is the percentage of light reflected from the polished surface of the mineral, and where possible values are given for each crystallographic orientation. The eye is not good at estimating absolute reflectance but is a good comparator. The reflectance values of the minerals should therefore be used for the purpose of comparing minerals. Reflectance can be related to a grey scale of brightness in the following way, but although followed in this book it is not a rigid scale. A mineral of reflectance $\sim 15\%$ (e.g. sphalerite) may appear to be light grey or white compared with a low reflectance mineral (such as quartz) or dark grey compared with a bright mineral (such as pyrite):

$R(\%)$	Grey scale
0–10	dark grey
10–20	grey
20–40	light grey
40–60	white
60–100	bright white

Bireflectance
This is a quantitative value, and for an anisotropic grain is a measure of the difference between the maximum and minimum reflectance values. However, bireflectance is usually assessed qualitatively, e.g.

Weak bireflectance: observed with difficulty, $\Delta R < 5\%$ (e.g. hematite)
Distinct bireflectance: easily observed, $\Delta R > 5\%$ (e.g. stibnite)

Pleochroism and bireflectance are closely related properties; the term pleochroism is used to describe change in tint or colour intensity, whereas bireflectance is used for a change in brightness.

1.6.2 Properties observed using crossed polars

The analyser is inserted into the optical path to give a dark image.

Anisotropy
This property varies markedly with crystallographic orientation of a section of a non-cubic mineral. Anisotropy is assessed as follows:

(a) Isotropic mineral: all grains remain dark on rotation of the stage, e.g. magnetite.
(b) Weakly anisotropic mineral: slight change on rotation, only seen on careful examination using slightly uncrossed polars, e.g. ilmenite.
(c) Strongly anisotropic mineral: pronounced change in brightness and possible colour seen on rotating the stage when using exactly crossed polars, e.g. hematite.

Remember that some cubic minerals (e.g. pyrite) can appear to be anisotropic, and weakly anisotropic minerals (e.g. chalcopyrite) may appear to be isotropic. Anisotropy and bireflectance are related properties; an anistropic grain is necessarily bireflecting, but the bireflectance in PPL is always much more difficult to detect than the anisotropy in crossed polars.

SYSTEMATIC DESCRIPTION OF MINERALS

Internal reflections

Light may pass through the polished surface of a mineral and be reflected back from below. Internal reflections are therefore shown by all transparent minerals. When one is looking for internal reflections, particular care should be paid to minerals of low to moderate reflectance (semi-opaque minerals), for which internal reflections might only be detected with difficulty and only near grain boundaries or fractures. Cinnabar, unlike hematite which is otherwise similar, shows spectacular red internal reflections.

1.6.3 The external nature of grains

Minerals have their grain shapes determined by complex variables acting during deposition and crystallisation and subsequent recrystallisation, replacement or alteration. Idiomorphic (a term used by reflected-light microscopists for well shaped or euhedral) grains are unusual, but some minerals in a polished section will be found to have a greater tendency towards a regular grain shape than others. In the ore mineral descriptions in Chapter 3, the information given under the heading 'crystals' is intended to be an aid to recognising minerals on the basis of grain shape. Textural relationships are sometimes also given.

1.6.4 Internal properties of grains

Twinning

This is best observed using crossed polars, and is recognised when areas with differing extinction orientations have planar contacts within a single grain. Cassiterite is commonly twinned.

Cleavage

This is more difficult to observe in reflected light than transmitted light, and is usually indicated by discontinuous alignments of regularly shaped or rounded pits. Galena is characterised by its triangular cleavage pits. Scratches sometimes resemble cleavage traces. Further information on twinning and cleavage is given under the heading of 'crystals' in the descriptions in Chapter 3.

Zoning

Compositional zoning of chemically complex minerals such as tetrahedrite is probably very common but rarely gives observable effects such as colour banding. Zoning of micro-inclusions is more common.

Inclusions

The identity and nature of inclusions commonly observed in the mineral is given, as this knowledge can be an aid to identification. Pyrrhotite, for example, often contains lamellar inclusions of pentlandite.

1.6.5 Vickers hardness number (VHN)

This is a quantitative value of hardness which is useful to know when comparing the polishing properties of minerals (see Section 1.9).

1.6.6 Distinguishing features

These are given for the mineral compared with other minerals of similar appearance. The terms harder or softer refer to comparative polishing hardness (see Section 1.8).

1.7 Observations using oil immersion in reflected-light studies

Preliminary observations on polished sections are always made simply with air (RI = 1.0) between the polished surface and the microscope objective, and for most purposes this suffices. However, an increase in useful magnification and resolution can be achieved by using immersion objectives which require oil (use microscope manufacturer's recommended oil, e.g. Cargille oil type A) between the objective lens and the section surface. A marked decrease in glare is also obtained with the use of immersion objectives. A further reason for using oil immersion is that the ensuing change in appearance of a mineral may aid its identification. Ramdohr (1969) states: 'It has to be emphasised over and over again that whoever shuns the use of oil immersion misses an important diagnostic tool and will never see hundreds of details described in this book.'

Table 1.1 The relationship between the reflectances of minerals in air (R_{air}) and oil immersion (R_{oil}) and their optical constants, refractive index (n) and absorption coefficient (k). Hematite is the only non-cubic mineral represented, and two sets of values corresponding to the ordinary (o) and extraordinary (e) rays are given. N is the refractive index of the immersion medium.

	n	k	$R_{air}(\%)$ ($N = 1.0$)	$R_{oil}(\%)$ ($N = 1.52$)
Transparent minerals				
fluorite CaF_2	1.434	0.0	3.2	0.08
sphalerite ZnS	2.38	0.0	16.7	4.9
Weakly absorbing minerals				
hematite Fe_2O_3 (o)	3.15	0.42	27.6	12.9
hematite Fe_2O_3 (e)	2.87	0.32	23.9	9.9
Absorbing (opaque) minerals				
galena PbS	4.3	1.7	44.5	28.9
silver Ag	0.18	3.65	95.1	93.2

Oil immersion nearly always results in a decrease in reflectance (Table 1.1), the reason being evident from examination of the Fresnel equation (Section 5.1.1), which relates the reflectance of a mineral to its optical properties *and* the refractive index (N) of the immersion medium. Because it is the $n-N$ and the $n+N$ values in the equation that are affected, the decrease in reflectance that results from the increase in N is greater for minerals with a lower absorption coefficient (see Table 1.1).

The colour of a mineral may remain similar or change markedly from air to oil immersion. The classic example of this is covellite, which changes from blue in air to red in oil, whereas the very similar blaubleibender covellite remains blue in both air and oil. Other properties, such as bireflectance and anisotropy, may be enhanced or diminished by use of oil immersion.

To use oil immersion, lower the microscope stage so that the immersion objective is well above the area of interest on the well levelled polished section. Place a droplet of recommended oil on the section surface and preferably also on the objective lens. Slowly raise the stage using the coarse focus control, viewing from the side, until the two droplets of oil just coalesce. Continue to raise the stage very slowly using the fine focus, looking down the eyepiece until the image comes into focus. Small bubbles may drift across the field but they should not cause any inconvenience. Larger bubbles, which tend to be caused by moving the sample too quickly, may only be satisfactorily removed by complete cleaning.

To clean the objective, lower the stage and immediately wipe the end of the objective with a soft tissue. Alcohol may be used with a tissue, but not a solvent such as acetone, which may result in loosening of the objective lens. The polished section can be carefully lifted from the stage and cleaned in the same way.

Most aspects of qualitative ore microscopy can be undertaken without resource to oil immersion, and oil immersion examination of sections which are subsequently to be carbon coated for electron beam microanalysis should be avoided. The technique is most profitably employed in the study of small grains of low reflectance materials such as graphite or organic compounds, where the benefits are a marked increase in resolution and image quality at high magnification.

1.8 Polishing hardness

During the polishing process, polished sections inevitably develop some relief (or topography) owing to the differing hardness of the component minerals. Soft minerals tend to be removed more easily than hard minerals. Also the surfaces of hard grains tend to become convex, whereas the surfaces of soft grains tend to become concave. One of the challenges of the polishing technique has been to totally avoid relief during polishing. This is because of the detrimental effect of polishing relief on the appearance of the polished section, as well as the necessity for optically flat polished surfaces for reflectance measurements. As some polishing relief is advantageous in *qualitative* mineral identification it is often beneficial to enhance the polishing relief by buffing the specimen for a few minutes using a mild abrasive such as gamma alumina on a soft nap.

Polishing relief results in a phenomenon known as the Kalb light line, which is similar in appearance to a Becke line. A sharp grain contact between a hard mineral such as pyrite and a soft mineral such as chalcopyrite should appear as a thin dark line when the specimen is exactly in focus. On defocusing slightly by increasing the distance between the specimen and objective, a fine line of bright light should appear along the grain contact in the softer mineral. The origin of this light line should easily be understood on examination of Figure 1.8. Ideally the light line should move away from the grain boundary as the specimen is further defocused. On defocusing in the opposite sense the light line appears in the harder mineral, and defocusing in this sense is often necessary as the white line is difficult to see in a bright white soft mineral. The light line is best seen using low power magnification and an almost closed aperture diaphragm.

The Kalb light line is used to determine the relative polishing hardness of minerals in contact in the same polished section. This sequence can be used to confirm optical identification of the mineral set, or as an aid to the identification of individual minerals, by comparison with published lists of relative polishing hardness (e.g. Uytenbogaardt & Burke 1971).

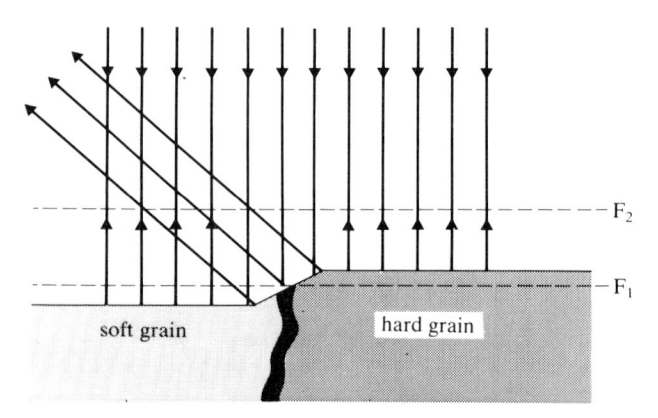

Figure 1.8 Relative polishing hardness. The position of focus is first at F_1. If the specimen is now *lowered* away from the objective, the level that is in focus will move to F_2, so that a light line (the 'Kalb light line') appears to move into the *softer* substance.

Relative polishing hardness can be of value in the study of microinclusions in an identified host phase; comparison of the hardness of an inclusion and its surround may be used to estimate the hardness of the inclusion or eliminate some of several possibilities resulting from identification attempted using optical properties. Similarly, if optical properties cannot be used to identify a mineral with certainty, comparison of polishing hardness with an identified coexisting mineral may help. For example, pyrrhotite is easily identified and may be associated with pyrite or pentlandite, which are similar in appearance; however, pyrite is harder than pyrrhotite whereas pentlandite is softer.

1.9 Microhardness (VHN)

The determination of relative polishing hardness (Section 1.8) is used in the mineral identification chart (Appendix C). Hardness can however be measured quantitatively using microindentation techniques. The frequently used hardness value, the Vickers hardness number (VHN), is given for each mineral listed in Appendix C.

Microindentation hardness is the most accurate method of hardness determination and, in the case of the Vickers technique, involves pressing a small square based pyramid of diamond into the polished surface. The diamond may be mounted in the centre of a special objective, with bellows enabling the load to be applied pneumatically (Fig. 1.9). The Commission on Ore Microscopy (COM) recommend

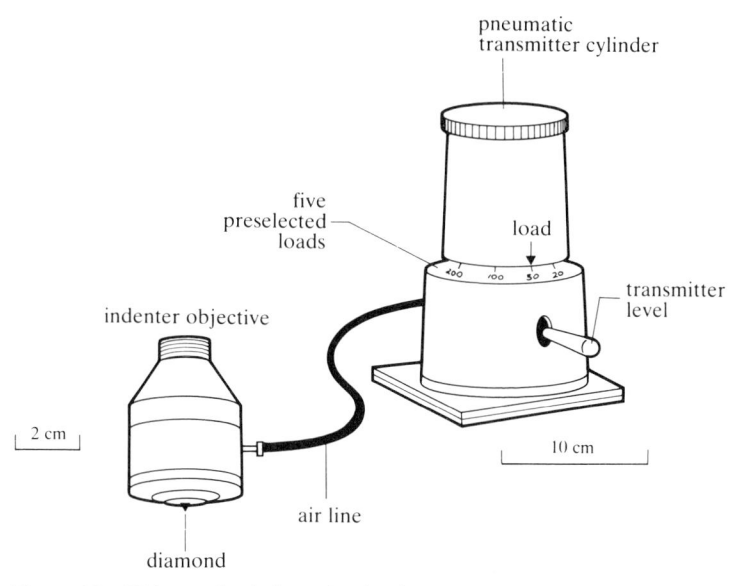

Figure 1.9 Vickers microindentation hardness tester.

that a load of 100 grams should be applied for 15 seconds. The size of the resulting square shaped impression depends on the hardness of the mineral:

$$\text{VHN} = \frac{1854 \times \text{load}}{d^2} \text{ kg/mm}^2$$

where the load is in kilograms and d is the average length of the diagonals of the impression in microns.

Hardness is expressed in units of pressure, that is, force per unit area. Thus the microindentation hardness of pyrite is written:

$$\text{pyrite, VHN}_{100} = 1027\text{--}1240 \text{ kg/mm}^2$$

The subscript 100 may be omitted as this is the standard load. As VHN values are always given in kg/mm² this is also often omitted.

The determination of hardness is a relatively imprecise technique, so an average of several indentations should be used. Tables of VHN usually give a range in value for a mineral, taking into account variations due to compositional variations, anistropy of hardness and uncertainty. Brittleness, plasticity and elasticity control the shape of the indentations, and as the shape can be useful in identification the COM recommends that indentation shape (using the abbreviations given in Fig. 1.10) be given with VHN values.

There is a reasonable correlation between VHN and Moh's scratch hardness as shown in Table 1.2.

1.10 Points on the use of the microscope (transmitted and reflected light)

Always focus using low power first. It is safer to start with the specimen surface close to the objective and *lower* the stage or raise the tube to achieve the position of focus.

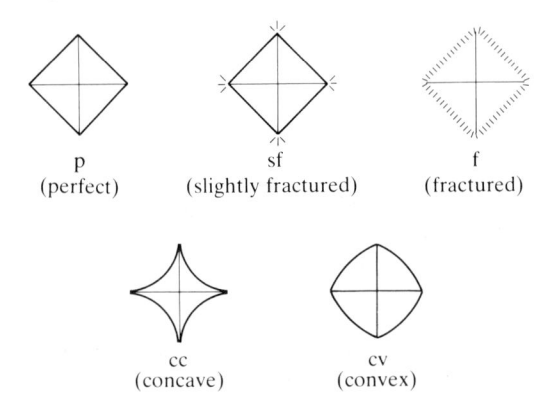

Figure 1.10 Indentation shapes.

p (perfect) sf (slightly fractured) f (fractured)

cc (concave) cv (convex)

POINTS ON USE OF MICROSCOPE

Table 1.2 Relation between VHN and Moh's hardness.

Moh's hardness (H)		~VHN
1	talc	10
2	gypsum	40
3	calcite	100
4	fluorite	200
5	apatite	500
6	orthoclase	750
7	quartz	1300
8	topaz	1700
9	corundum	2400
[10	diamond]	

Thin sections must always be placed on the stage with the cover slip on top of the section, otherwise high power objectives may not focus properly.

Polished samples must be level. Blocks may be mounted on a small sphere of plasticine on a glass plate and pressed gently with a levelling device. Carefully machined polished blocks with parallel faces can usually be placed directly on the stage. A level sample should appear uniformly illuminated. A more exact test is to focus on the samples, then close the aperture diaphragm (seen using the Bertrand lens) and rotate the stage. The small spot of light seen as the image should not wobble if the sample is level.

Good polished surfaces require careful preparation and are easily ruined. Never touch the polished surface or wipe it with anything other than a clean soft tissue, preferably moistened with alcohol or a proprietary cleaning fluid. Even a dry tissue can scratch some soft minerals. Specimens not in use should be kept covered or face down on a tissue.

The analyser is usually fixed in orientation on transmitted-light microscopes but the polariser may be free to rotate. There is no need to rotate the polariser during normal use of the microscope and it should be positioned to give east–west vibrating polarised light. To check that the polars are exactly crossed examine an isotropic substance such as glass and adjust the polariser to give maximum darkness (complete extinction).

The alignment of polariser and analyser for reflected light can be set approximately fairly easily. Begin by obtaining a level section of a bright isotropic mineral such as pyrite. Rotate the analyser and polariser to their zero positions, which should be marked on the microscope. Check that the polars are crossed, i.e. the grain is dark. Rotate the analyser slightly to give as dark a field as possible. View the polarisation figure (see Section 1.4). Adjust the analyser (and/or polariser) until a perfectly centred black cross is obtained. Examine an optically homogeneous area of a uniaxial mineral such as ilmenite, niccolite or hematite. Using

THE MICROSCOPIC STUDY OF MINERALS

crossed polars it should have four extinction positions at 90°, and the polarisation colours seen in each quadrant should be identical. Adjust the polariser and analyser until the best results are obtained (see Hallimond 1970, p. 101).

Ensure that the stage is well centred using the high power objective before studying optical figures.

1.11 Thin- and polished-section preparation

Thin sections are prepared by cementing thin slices of rock to glass and carefully grinding using carborundum grit to produce a paper thin layer of rock. The standard thickness of 30 microns is estimated using the interference colours of known minerals in the section. A cover slip is finally cemented on top of the layer of rock (Fig. 1.11).

The three common types of polished section are shown in Figure 1.11. Preparation of a polished surface of a rock or ore sample is a rather involved process which involves five stages:

(1) *Cutting* the sample with a diamond saw.
(2) *Mounting* the sample on glass or in a cold-setting resin.

Figure 1.11 Sections.

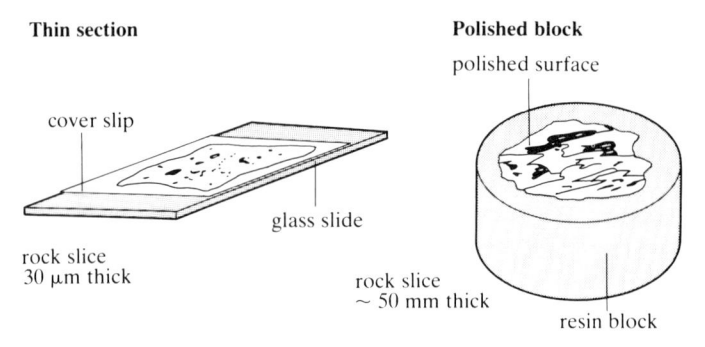

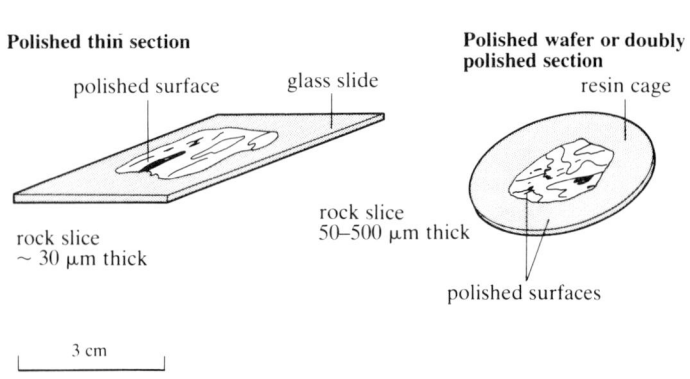

(3) *Grinding* the surface flat using carborundum grit and water on a glass or a metal surface.
(4) *Polishing* the surface using diamond grit and an oily lubricant on a relatively hard 'paper' lap.
(5) *Buffing* the surface using gamma alumina powder and water as lubricant on a relatively soft 'cloth' lap.

There are many variants of this procedure, and the details usually depend on the nature of the samples and the polishing materials, and equipment that happen to be available. Whatever the method used, the objective is a flat, relief-free, scratch-free polished surface. The technique used by the British Geological Survey is outlined by B. Lister (1978).

Polished wafers are used in the study of fluid inclusions. Polished thin sections with **both** sides polished are recommended for the study of minerals with high refractive indices, such as sphalerite and carbonates.

2 Silicate minerals

2.1 Crystal chemistry of silicate minerals

All silicate minerals contain silicate oxyanions $[SiO_4]^{4-}$. These units take the form of a tetrahedron, with four oxygen ions at the apices and a silicon ion at the centre. The classification of silicate minerals depends on the degree of polymerisation of these tetrahedral units. In silicate minerals, a system of classification commonly used by mineralogists depends upon how many oxygens in each tetrahedron are shared with other similar tetrahedra.

Nesosilicates
Some silicate minerals contain independent $[SiO_4]^{4-}$ tetrahedra. These minerals are known as nesosilicates, orthosilicates, or *island silicates*. The presence of $[SiO_4]$ units in a chemical formula of a mineral often indicates that it is a nesosilicate, e.g. olivine $(Mg,Fe)_2SiO_4$ or garnet $(Fe,Mg \text{ etc.})_3Al_2Si_3O_{12}$, which can be rewritten as $(Fe,Mg \text{ etc.})_3 Al_2[SiO_4]_3$. Nesosilicate minerals include the olivine group, the garnet group, the Al_2SiO_5 polymorphs (andalusite, kyanite, sillimanite), zircon, sphene, staurolite, chloritoid, topaz and humite group minerals.

Cyclosilicates
Cyclosilicates or *ring silicates* may result from tetrahedra sharing two oxygens, linked together to form a ring, whose general composition is $[Si_xO_{3x}]^{2x-}$, where x is any positive integer. The rings are linked together by cations such as Ba^{2+}, Ti^{4+}, Mg^{2+}, Fe^{2+}, Al^{3+} and Be^{2+}, and oxycomplexes such as $[BO_3]^{3-}$ may be included in the structure. A typical ring composition is $[Si_6O_{18}]^{12-}$ and cyclosilicates include tourmaline, cordierite and beryl, although cordierite and beryl may be included with the tektosilicates in some classifications.

Sorosilicates
Sorosilicates contain $[Si_2O_7]^{6-}$ groups of two tetrahedra sharing a common oxygen. The $[Si_2O_7]^{6-}$ groups may be linked together by Ca^{2+}, Al^{3+}, Mg^{2+}, Fe^{2+} and some rare earth ions (Ce^{2+}, La^{2+} etc.), and also contain $(OH)^-$ ions in the epidote group of minerals. Besides the epidote group, sorosilicates include the melilites, vesuvianite (or idocrase) and pumpellyite.

Inosilicates
When two or two and a half oxygens are shared by adjacent tetrahedra, inosilicates or *chain silicates* result. Minerals in this group are called

CRYSTAL CHEMISTRY

single chain silicates because the $[SiO_4]^{4-}$ tetrahedra are linked together to form chains of composition $[SiO_3]_n^{2-}$ stacked together parallel to the c axis, and bonded together by cations such as Mg^{2+}, Fe^{2+}, Ca^{2+} and Na^+ (Fig. 2.1). Chain silicate minerals always have a prismatic habit and exhibit two prismatic cleavages meeting at approximately right angles on the basal plane, these cleavages representing planes of weakness between chain units. The pyroxenes are single chain inosilicates. Variations in the structure of the single chain from the normal pyroxene structure produces a group of similar, though structurally different, minerals (called the pyroxenoids, of which wollastonite is a member).

Double chain silicates also exist in which double chains of composition $[Si_4O_{11}]_n^{6-}$ are stacked together, again parallel to the c crystallographic axis, and bonded together by cations such as Mg^{2+}, Fe^{2+}, Ca^{2+}, Na^+ and K^+ with $(OH)^-$ anions also entering the structure (Fig. 2.2). Double chain minerals are also prismatic and possess two prismatic cleavages meeting at approximately 126° on the basal plane, these cleavages again representing planes of weakness between the double chain units. The amphiboles are double chain inosilicates.

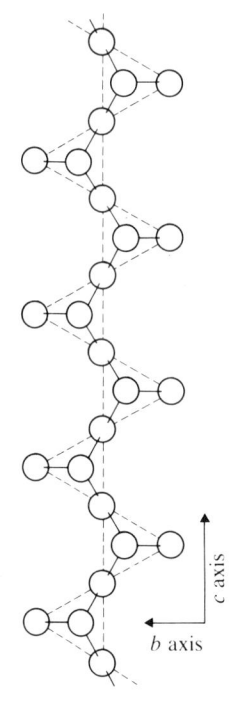

Figure 2.1 Single chain silicates.

single chain parallel to the c axis as occurs in the pyroxenes

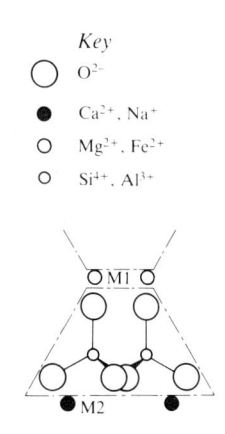

single chains viewed at right angles to the c axis: the chains are linked together by various atoms in the positions shown

Key
○ O^{2-}
● Ca^{2+}, Na^+
○ Mg^{2+}, Fe^{2+}
○ Si^{4+}, Al^{3+}

SILICATE MINERALS

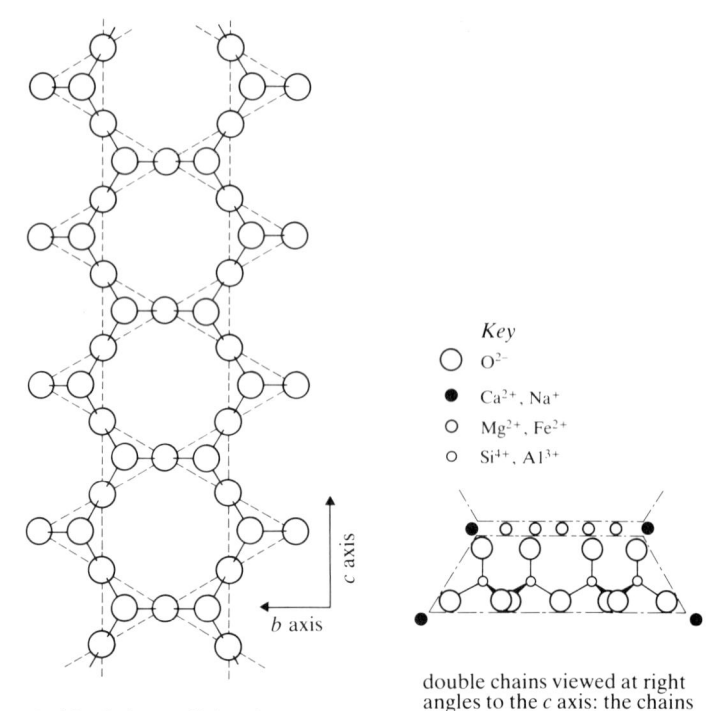

Figure 2.2 Double chain silicates.

double chain parallel to the c axis as occurs in the amphiboles

Key
○ O^{2-}
● Ca^{2+}, Na^+
○ Mg^{2+}, Fe^{2+}
○ Si^{4+}, Al^{3+}

double chains viewed at right angles to the c axis: the chains are linked together by various atoms in the positions shown

Phyllosilicates

When three oxygens are shared between tetrahedra, phyllosilicates or *sheet silicates* result. The composition of such a silicate sheet is $[Si_4O_{10}]_n^{4-}$. Phyllosilicates exhibit 'stacking', in which a sheet of brucite composition containing Mg^{2+}, Fe^{2+} and $(OH)^-$ ions, or a sheet of gibbsite composition containing Al^{3+} and $(OH)^-$ ions, is stacked on to an $[Si_4O_{10}]$ silicate sheet or sandwiched between two $[Si_4O_{10}]$ silicate sheets (Fig. 2.3a). Variations in this stacking process give rise to several related mineral types called *polytypes*. Three main polytypes exist, each of which is defined by the repeat distance of a complete multilayered unit measured along the crystallographic axis. The 7 Å, two layer structure includes the mineral kaolin; the 10 Å, three layer structure includes the clay minerals montmorillonite and illite, and also the micas; and the 14 Å, four layer structure includes chlorite. Figure 2.3b gives simplified details of the main polytypes. These multilayer structures are held together by weakly bonded cations (K^+, Na^+) in the micas and other 10 Å and 14 Å polytypes. In some other sheet silicates, only Van der Waals bonding occurs between these multilayer structures. The sheet silicates cleave easily along this weakly bonded layer, and all of them

(a) Idealised tetrahedral layer of the sheet silicates

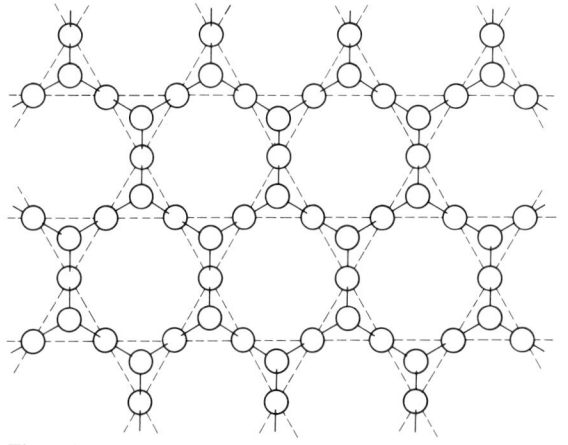

Figure 2.3
(a) Sheet silicates
(b) sheet silicates, the three polytypes.

The apices of the tetrahedra all point in the same direction (in this case upwards). Such a tetrahedral sheet may be depicted in cross section as:

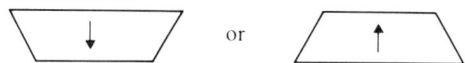

These Si-O layers are joined together by octahedral layers; either (Al-OH) layers, called Gibbsite layers and depicted by the letter G, or (Mg,Fe-OH) layers, called Brucite layers and depicted by the letter B.

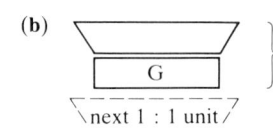

(b)

} 2 layer unit (1 tetrahedral layer and 1 octahedral layer; called a 1 : 1 type)

(1) 7 Å type represented by kaolinite – serpentine is similar with a B layer replacing the G layer

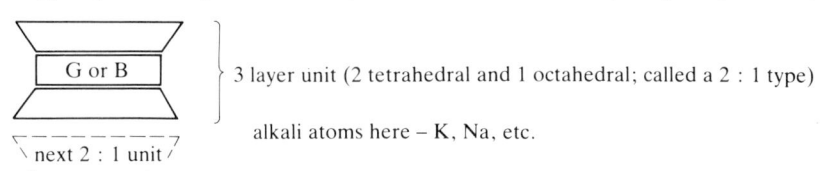

} 3 layer unit (2 tetrahedral and 1 octahedral; called a 2 : 1 type)

alkali atoms here – K, Na, etc.

(2) 10 Å type with muscovite, illite and montmorillonite having G octahedral layers, and biotite B layers: the three layer units are joined together by monovalent alkali ions. Montmorillonite may not possess any atoms in this plane and may have an overall negative charge. Water molecules may enter the structure along these inter-unit planes

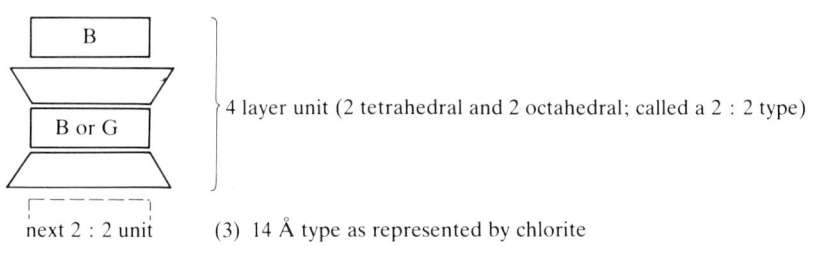

} 4 layer unit (2 tetrahedral and 2 octahedral; called a 2 : 2 type)

(3) 14 Å type as represented by chlorite

SILICATE MINERALS

exhibit this perfect cleavage parallel to the basal plane. Minerals belonging to this group include micas, clay minerals, chlorite, serpentine, talc and prehnite.

Tektosilicates
When all four oxygens are shared with other tetrahedra, tektosilicates or *framework silicates* form. Such a framework structure, if composed entirely of silicon and oxygen, will have the composition SiO_2 as in quartz. However, in many tektosilicates the silicon ion (Si^{4+}) is replaced by aluminium (Al^{3+}). Since the charges do not balance, a *coupled substitution* occurs. For example, in the alkali feldspars, one aluminium ion plus one sodium ion enter the framework structure and replace one silicon ion and, in addition, fill a vacant site. This can be written

$$Al^{3+} + Na^+ \rightleftharpoons Si^{4+} + \square \text{ (vacant site)}$$

In plagioclase feldspars a slightly different coupled substitution is required since the calcium ion is divalent:

$$2Al^{3+} + Ca^{2+} \rightleftharpoons 2Si^{4+} + \square \text{ (vacant site)}$$

This type of coupled substitution is common in the *feldspar* minerals, and more complex substitutions occur in other tektosilicate minerals or mineral groups. Tektosilicates include feldspars, quartz, the feldspathoid group, scapolite and the zeolite group.

The classification of each mineral or mineral group is given in the descriptions in Section 2.2.

2.2 Mineral descriptions

The thin-section information on the silicate minerals is laid out in the same way for each mineral as follows:

	Group	Crystal chemistry
Mineral name	Composition (note: Fe means Fe^{2+})	Crystal system
	Drawing of mineral (if needed)	

RI data
Birefringence (δ): Maximum birefringence is given for each mineral. Any variation quoted depends upon mineral composition.
Uniaxial or biaxial data with sign +ve (positive) or −ve (negative).
Specific gravity or density   Hardness

Then the main properties of each mineral are given in the following order: colour, pleochroism, habit, cleavage, relief, alteration, birefringence, interference figure, extinction angle, twinning and others (zoning etc.). Of course, only those properties which a particular mineral possesses are actually given, and the important properties are marked with an asterisk.

Some mineral descriptions may include a short paragraph on their distinguishing features and how the mineral can be recognised from other minerals with similar optical properties.

The description ends with a short paragraph on the mineral occurrences, associated minerals and the rocks in which it is found.

Al_2SiO_5 polymorphs

Nesosilicates

Andalusite Al_2SiO_5

orthorhombic
$0.983:1:0.704$

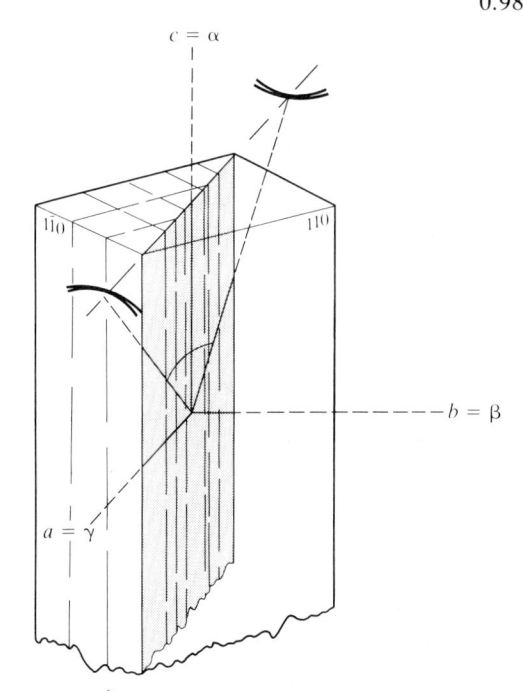

$n_\alpha = 1.629–1.649$
$n_\beta = 1.633–1.653$ } RI variation in all polymorphs is due to ferric iron and manganese entering structure
$n_\gamma = 1.638–1.660$
$\delta = 0.009–0.011$
$2V_\alpha = 78°–86°$ −ve (a prism section is length fast)
OAP is parallel to (010)
$D = 3.13–3.16 \quad H = 6½–7½$

COLOUR Colourless but may be weakly coloured in pinks.
PLEOCHROISM Rare but some sections show α pink, β and γ greenish yellow.
*HABIT Commonly occurs as euhedral elongate prisms in metamorphic rocks which have suffered medium grade thermal metamorphism (var. chiastolite). Prisms have a square cross section (a basal section is square).
*CLEAVAGE $\{110\}$ good appearing as traces parallel to the prism edge in prismatic sections but intersecting at right angles in a basal section.
RELIEF Moderate.
ALTERATION Andalusite can invert or change to sillimanite with increasing metamorphic grade. Under hydrothermal conditions or retrograde metamorphism andalusite changes to sericite (a type of muscovite), thus:

$$\underset{}{3Al_2SiO_5} + 2H_2O + \underset{\text{from feldspar}}{(3SiO_2 + K_2O)} \rightarrow \underset{\text{sericite}}{K_2Al_4Si_6Al_2(OH)_4O_{20}}$$

*BIREFRINGENCE Low, first order (similar to quartz).
*EXTINCTION Straight on prism edge or on $\{110\}$ cleavages.
INTERFERENCE FIGURE Basal section gives a Bx_a figure but $2V$ is too large to see in field of view. Look for an isotropic section approx. (101), and obtain an optic axis figure which will be negative.
OTHER FEATURES Crystals in metamorphic rocks are usually poikiloblastic, and full of quartz inclusions.
*OCCURRENCE See after sillimanite.

Kyanite Al$_2$SiO$_5$
triclinic
0.902 : 1 : 0.710
$\alpha = 90°5'$, $\beta = 101°2'$, $\gamma = 105°44'$

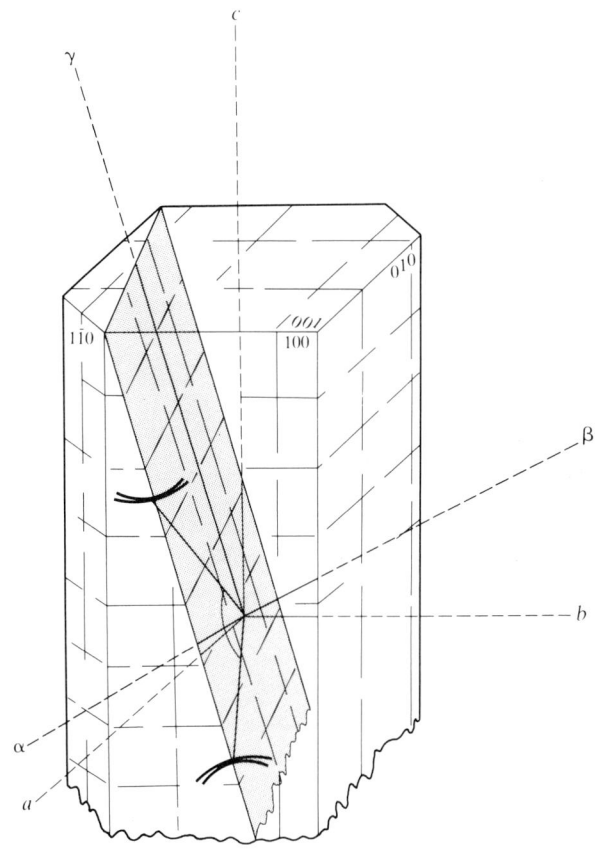

n_α = 1.712–1.718
n_β = 1.721–1.723
n_γ = 1.727–1.734
δ = 0.015–0.016
$2V_\alpha = 82°$ −ve
OAP is approx. perpendicular to (100) with the a axis approximately the acute bisectrix
D = 3.58–3.65 H = 5½–7

COLOUR Usually colourless in thin section but may be pale blue.
PLEOCHROISM Weak but seen in thick sections with α colourless, β and γ blue.
HABIT Usually found as subhedral prisms in metamorphic rocks. The prisms are blade-like, i.e. broad in one direction but thin in a direction at right angles to this.

SILICATE MINERALS

*CLEAVAGE	$\{100\}$ and $\{010\}$ very good. Parting present on $\{001\}$.
*RELIEF	High: the high relief, which is easily seen if the section is held up to the light, is a very distinctive feature.
ALTERATION	As andalusite. Kyanite often occurs within large 'knots' of micaceous minerals; it also inverts to sillimanite with increasing temperature.
BIREFRINGENCE	Low.
*EXTINCTION	Oblique on cleavages and prism edge; $\gamma\hat{\,}$ prism edge is ~30°.
INTERFERENCE FIGURE	(100) sections give Bx_a figures; but as with andalusite an isotropic section should be obtained and a single isogyre used to obtain sign and size of $2V$.
TWINNING	Multiple twinning occurs on $\{100\}$.
OTHER FEATURES	The higher birefringence and excellent $\{100\}$ cleavage, intersected by the $\{001\}$ parting on the prism face, help to distinguish kyanite from andalusite and other index minerals.
*OCCURRENCE	See after sillimanite.

Al$_2$SiO$_5$ POLYMORPHS

Sillimanite Al$_2$SiO$_5$ orthorhombic
 0.980 : 1 : 0.757

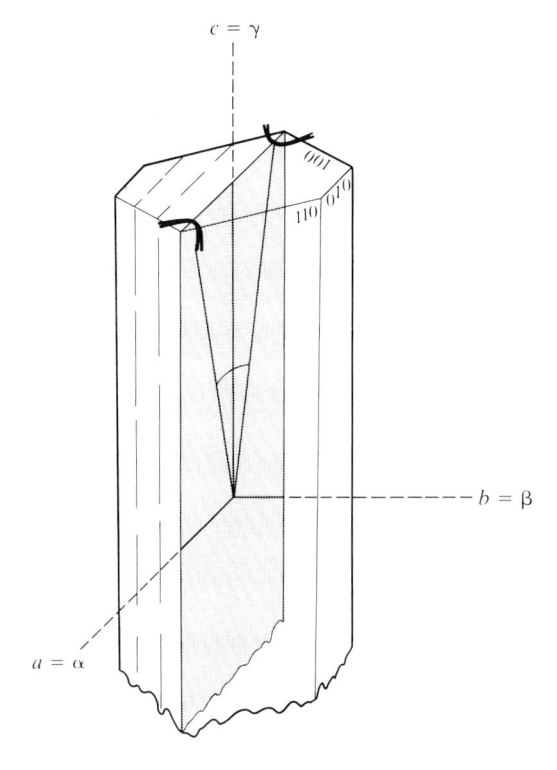

n_α = 1.654–1.661
n_β = 1.658–1.662
n_γ = 1.678–1.683
δ = 0.019–0.022
$2V_\gamma$ = 21°–30° +ve (a prism section is length slow)
OAP is parallel to (010)
D = 3.23–3.27 H = 6½–7½

COLOUR Colourless.
*HABIT It occurs as elongate prisms in two habits: either as small fibrous crystals found in regionally metamorphosed schists and gneisses, or as small prismatic crystals growing from andalusite in thermal aureoles.
*CLEAVAGE {010} perfect: thus a basal section of sillimanite, which is diamond shaped, has cleavages parallel to the long axis.
RELIEF Moderate.
ALTERATION Rare.
*BIREFRINGENCE Moderate.
*EXTINCTION Straight on single cleavage trace.

SILICATE MINERALS

*INTERFERENCE FIGURE
Basal section gives an excellent Bx_a (+ve) figure with a small $2V$. Note that basal sections are usually small, so a very high power objective lens will give the best figure (×55 or more).

*OTHER FEATURES
In high grade regionally metamorphosed rocks the fibrous sillimanite (formerly called fibrolite) is usually found associated with biotite, appearing as long thin fibres growing within the mica crystal.

*OCCURRENCE
All three polymorphs can be used as index minerals in metamorphic rocks. They all develop in alumina-rich pelites under different conditions of temperature and pressure (Fig. 2.4.). Andalusite forms at low pressures (< 1.5 kb) and low to moderate temperatures in thermal aureoles and regional metamorphism of Buchan type (high heat flow, low P). At higher temperatures it inverts to sillimanite. Kyanite forms at medium to high pressures and low to moderate temperatures in regional metamorphism of Barrovian type (high heat flow, moderate or high P). At higher temperatures kyanite also inverts to sillimanite which occurs over a wide range of pressures and high temperatures. The sequences of mineralogical changes in pelites are:

(a) Buchan (low P, high heat flow ~60 °C/km): (low grade) micas – andalusite (+ cordierite) – sillimanite (high grade).
(b) Barrovian (moderate to high P, high heat flow ~30 °C/km): (low grade) micas – staurolite – garnet – kyanite – sillimanite (highest grade).

The P–T diagram (Fig. 2.4) shows the stability relations of the three polymorphs. The minimum melting curve of granite has been superimposed on to the diagram. To the right (up temperature) side of this curve melting has taken place and the polymorphs would therefore occur in metamorphic rocks which had undergone some melting (e.g. migmatitic rocks).

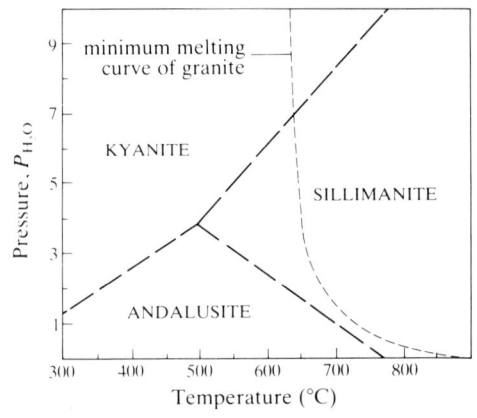

Figure 2.4 Stability relations of the three Al_2SiO_5 polymorphs. Also shown is the melting curve for albite + orthoclase + quartz + water, representing granite.

Sillimanite can also occur in high temperature xenoliths found as residual products in aluminous rocks after partial melting has taken place. All the Al_2SiO_5 polymorphs have been recognised as detrital minerals in sedimentary rocks.

Amphibole group   Inosilicates

Introduction

The amphiboles include orthorhombic and monoclinic minerals. They possess a double chain silicate structure which allows a large number of elemental substitutions. The double chain has a composition of $(Si_4O_{11})_n$, with some substitution by Al^{3+} for silicon. The chains are joined together by ions occupying various sites within the structure, and these sites are called A, X and Y. The Y sites are usually occupied by Mg^{2+} and Fe^{2+}, although Fe^{3+}, Al^{3+}, Mn^{2+} and Ti^{4+} may also enter the Y sites. The X sites are usually filled by Ca^{2+} or Ca^{2+} and Na^+, although the orthorhombic amphiboles have Mg^{2+} or Fe^{2+} occupying the X sites as well as the Y ones. The A sites are always occupied by Na^+, although in the calcium-poor and calcium-rich amphiboles the A sites usually remain unoccupied.

The main amphibole groups include:

(a) The Ca-poor amphiboles (Ca + Na nearly zero), which include the orthorhombic amphiboles and the Ca-poor monoclinic amphiboles. The minerals included are the anthophyllite–gedrite group (which have properties extremely similar to the cummingtonite–grunerite group in the monoclinic amphiboles). The general formula is:

$$X_2Y_5Z_8O_{22}(OH,F)_2$$

where $X = Mg,Fe$, $Y = Mg,Fe,Al$ and $Z = Si,Al$.

(b) The Ca-rich amphiboles (with Ca > Na) are monoclinic, and include the common hornblendes and tremolite–ferroactinolite. The general formula is:

$$AX_2Y_5Z_8O_{22}(OH,F)_2$$

with $A = Na$ (or zero in some members), $X = Ca$, $Y = Mg,Fe,Al$ and $Z = Si,Al$.

(c) The alkali amphiboles are also monoclinic (with Na > Ca), and the general formula is:

$$AX_2Y_5Z_8O_{22}(OH,F)_2$$

where $A = Na$, $X = Na$ (or Na,Ca), $Y = Mg,Fe,Al$ and $Z = Si,Al$. The main members are glaucophane–riebeckite, richterite and eckermannite–arfvedsonite.

SILICATE MINERALS

The amphiboles will be examined in the order above, i.e. subgroups (a), (b) and (c), but the general optical properties of all amphibole minerals are given below:

COLOUR PLEOCHROISM
: Green, yellow and brown in pale or strong colours. Mg-rich amphiboles may be colourless or possess pale colours with slight pleochroism, whereas iron-rich and alkali amphiboles usually are strongly coloured and pleochroic.

HABIT
: Amphiboles usually occur as elongate prismatic minerals, often with diamond shaped cross sections.

*CLEAVAGE
: All amphiboles have two prismatic cleavages which intersect at 56° (acute angle).

RELIEF
: Moderate to high.

ALTERATION
: Common in all amphiboles; usually to chlorite or talc in the presence of water. A typical reaction is as follows:

$$Mg_2Mg_5Si_8O_{22}(OH, F)_2 + H_2O \rightarrow Mg_6Si_8O_{20}(OH)_4 + Mg(OH)_2$$
Mg anthophyllite talc brucite

BIREFRINGENCE
: Low to moderate; upper first order or lower second order interference colours occur, iron-rich varieties always giving higher interference colours. The strong colours of alkali amphiboles often mask their interference colours.

INTERFERENCE FIGURE
: Apart from glaucophane and katophorite, most amphiboles have large $2V$ angles; thus an isotropic section is needed to examine a single optic axis figure. In the alkali amphiboles dispersion is so strong that interference figures may not be seen.

EXTINCTION
: Orthorhombic amphiboles have parallel (straight) extinction. All other amphiboles are monoclinic with variable maximum extinction angles (Fig. 2.5).

ZONING
: Fairly common.

TWINNING
: Common on $\{100\}$; with either single or multiple twins present.

AMPHIBOLE GROUP

Ca-poor amphiboles

Anthophyllite $(Mg,Fe)_2(Mg,Fe)_5Si_8O_{22}(OH,F)_2$ .... $Mg \gg Fe$ } orthorhombic
Gedrite $(Mg,Fe)_2(Mg,Fe)_3Al_2(Si_6Al_2)O_{22}(OH,F)_2$ .. $Fe \gg Mg$ }

0.967 : 1 : 0.285

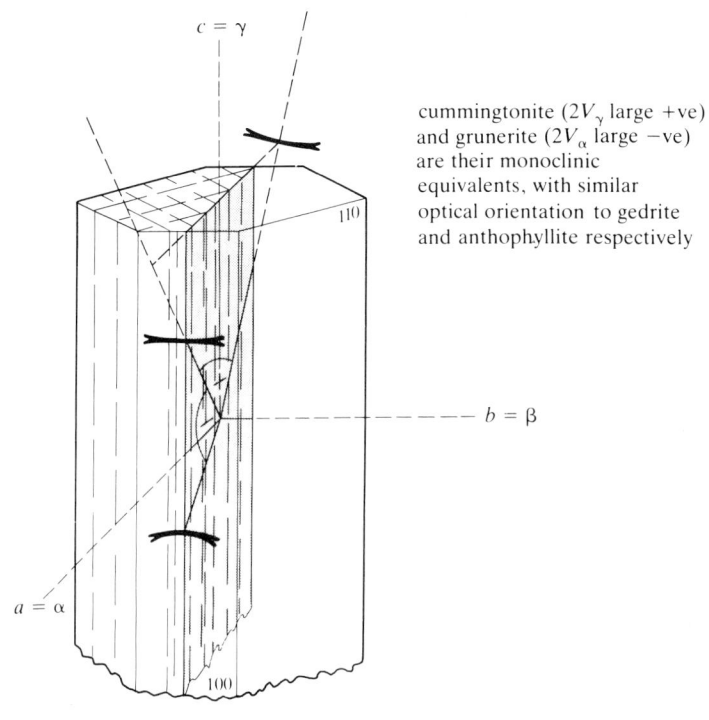

cummingtonite ($2V_\gamma$ large +ve) and grunerite ($2V_\alpha$ large −ve) are their monoclinic equivalents, with similar optical orientation to gedrite and anthophyllite respectively

$c = \gamma$

$b = \beta$

$a = \alpha$

n_α = 1.596–1.694
n_β = 1.605–1.710
n_γ = 1.615–1.722
δ = 0.013–0.028
$2V_\alpha$ = 69°–90° (anthophyllite) −ve }
$2V_\gamma$ = 78°–90° (gedrite) +ve } both crystals are length slow
OAP parallel to (010)
D = 2.85–3.57 H = 5½–6

COLOUR   Pale brown to pale yellow.
*PLEOCHROISM Gedrite has a stronger pleochroism than anthophyllite with α and β pale brown, γ darker brown.
HABIT   Elongate prismatic crystals; basal sections recognised by intersecting cleavages.
*CLEAVAGE   Two prismatic {110} cleavages intersecting at 54° (126°). The two cleavages are parallel to each other in a prism section and so elongate prismatic sections appear to have only one cleavage.

43

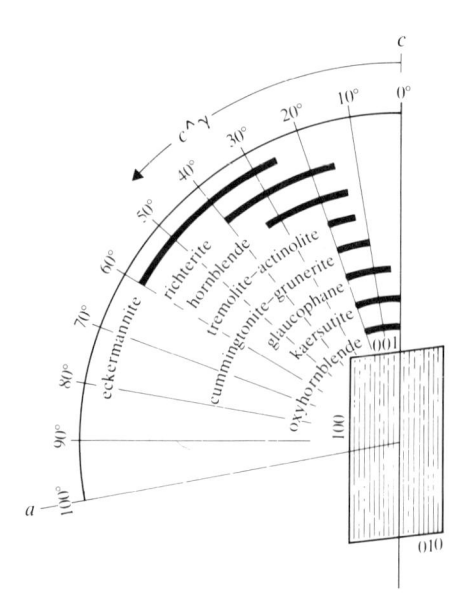

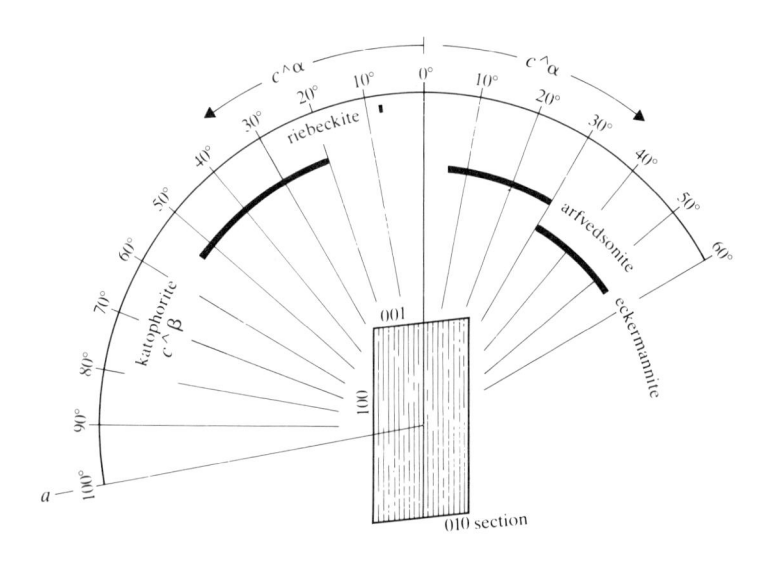

Figure 2.5 Extinction angles of amphiboles. Note that $c\hat{}\beta$ for katophorite will be cleavage$\hat{}$slow ray, since the other component in this orientation is α.

AMPHIBOLE GROUP

RELIEF	Moderate.
*ALTERATION	Common (see introduction).
BIREFRINGENCE	Low to moderate.
INTERFERENCE FIGURE	Bx_a figure seen on a (100) prismatic face (anthophyllite) or a basal face (gedrite) but crystals are usually so small that figures may be impossible to obtain. Best results will be obtained from a single optic axis figure.
EXTINCTION	Straight; crystals are length slow.
OCCURRENCE	Unknown in igneous rocks, the orthorhombic amphiboles occur widely in metamorphic rocks, with anthophyllite found in association with cordierite.

Cummingtonite **Grunerite**   Cummingtonite and grunerite are the monoclinic equivalents of anthophyllite and gedrite. Cummingtonite (the Mg-rich form) is positive, whereas grunerite (the Fe-rich form) is negative. $2V$ is large, and density and hardness are similar to anthophyllite–gedrite. Birefringence is moderate to high (grunerite) and each mineral has oblique extinction with $\gamma\hat{}\,$cleavage $= 10°$ to $21°$ on the (010) prism face (see Fig. 2.5).

Cummingtonite occurs in metamorphosed basic igneous rocks, where it is associated with common hornblendes. Grunerite occurs in metamorphosed iron-rich sediments, where it is associated with either magnetite and quartz or with almandine garnet and fayalitic olivine, the latter minerals being common constituents of eulysite bands.

Amosite (brown asbestos) is asbestiform grunerite.

SILICATE MINERALS

Ca-rich amphiboles

Tremolite–ferroactinolite $Ca_2(Mg,Fe)_5Si_8O_{22}(OH,F)_2$ monoclinic
$0.55:1:0.29, \beta = 105°$

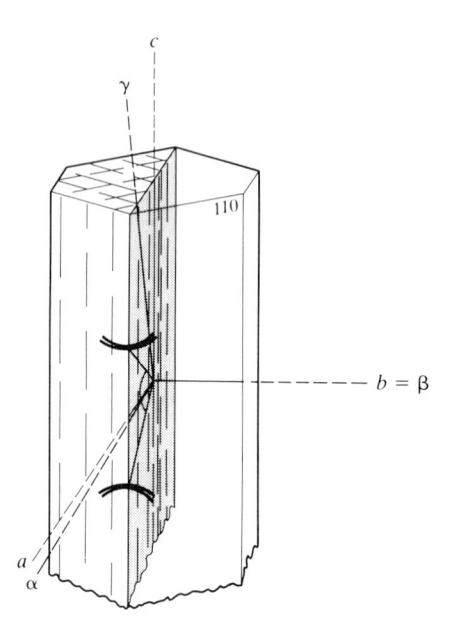

$n_\alpha = 1.599-1.688$
$n_\beta = 1.612-1.697$
$n_\gamma = 1.622-1.705$
$\delta = 0.027-0.017$
$2V_\alpha = 86°-65°$ −ve
OAP is parallel to (010)
$D = 3.02-3.44$ $H = 5-6$

COLOUR — Colourless to pale green (tremolite). Ferroactinolite is pleochroic in shades of green.
*PLEOCHROISM — Related to iron content – the more iron rich, the more pleochroic the mineral, with α pale yellow, β yellowish green, γ greenish blue.
HABIT — Elongate prismatic with aggregates of fibrous crystals also present.
*CLEAVAGE — The usual prismatic cleavages $\{110\}$ and intersecting at 56° on the basal plane.
RELIEF — Moderate to high.
ALTERATION — Common (see introduction).
BIREFRINGENCE — Moderate: second order green is maximum interference colour seen on a prismatic section parallel to (010).
INTERFERENCE FIGURE — Large 2V seen on (100) prismatic section. It is best to find an isotropic section, examine one optic axis and get sign and size of 2V.

AMPHIBOLE GROUP

*EXTINCTION ANGLE The extinction angle of slow cleavage varies from 21° to 11° depending upon the Mg:Fe ratio; the higher the ratio, the higher the extinction angle. Thus $\gamma \hat{} cl = 21°$ to 17° in tremolite and 17° to 11° in ferroactinolite. Most amphiboles are nearly length slow. In most thin sections, the prismatic section will rarely be correctly oriented to give a maximum extinction angle; for example the extinction will vary from straight on a (100) section to a maximum angle on an (010) section.

*TWINNING Amphiboles are frequently simply twinned with $\{100\}$ as twin plane. This is shown under crossed polars by a plane across the long axis of the basal section, splitting the section into two twin halves. Multiple twinning on $\{100\}$ may also occur.

OCCURRENCE Tremolite (and actinolite) are metamorphic minerals forming during both thermal and regional metamorphism, especially from impure dolomitic limestones. At high grades tremolite is unstable, breaking down in the presence of calcite to form diopside or in the presence of dolomite to give olivine. Tremolite–actinolite form during the metamorphism of ultrabasic rocks at low grades. Actinolite is a characteristic mineral of greenschist facies rocks, occurring with common hornblende, and may also occur in blueschist rocks in association with glaucophane, epidote, albite and other minerals. Amphibolisation (or, uralitisation) of basic igneous rocks is the name given to the alteration of pyroxene minerals to secondary amphibole by the pneumatolytic action of hydrous magmatic liquids on the igneous rocks, and the amphibole so formed may be a tremolite or actinolite.

Nephrite is the asbestiform variety of tremolite–actinolite. Precious jade is either nephrite or jadeite.

Hornblende series $Na_{0-1}Ca_2(Mg_{3-5}Al_{2-0})(Si_{6-7}Al_{2-1})O_{22}(OH,F)_2$
('Common' hornblende)

The hornblende series is the name given to amphiboles which define a 'field' of composition the boundary end-members of which are represented by the four phases:

hastingsite $Ca_2Mg_4Al(Si_7Al)O_{22}(OH,F)_2$
tschermakite $Ca_2Mg_3Al_2(Si_6Al_2)O_{22}(OH,F)_2$
edenite $NaCa_2Mg_5(Si_7Al)O_{22}(OH,F)_2$
pargasite $NaCa_2Mg_4Al(Si_6Al_2)O_{22}(OH,F)_2$

Iron (Fe^{2+}) may replace Mg in hornblendes but this has been omitted from the formulae for simplicity. The hornblende field can be represented in a graph by plotting the number of sodium atoms in the formulae against either the number of aluminium atoms replacing silicon or the number of aluminium atoms replacing magnesium (Fig. 2.6).

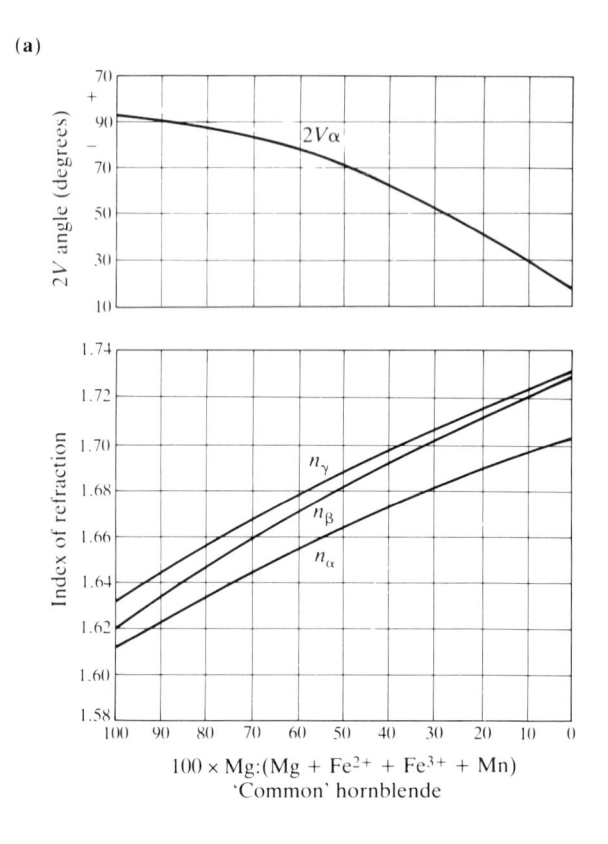

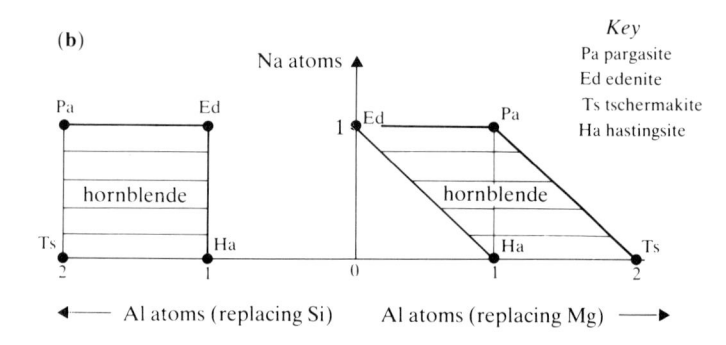

Figure 2.6 (a) Variation of 2V angle and indices of refraction in the 'common' hornblende series (after Deer, Howie & Zussman 1962) (b) field of common hornblende compositions.

AMPHIBOLE GROUP

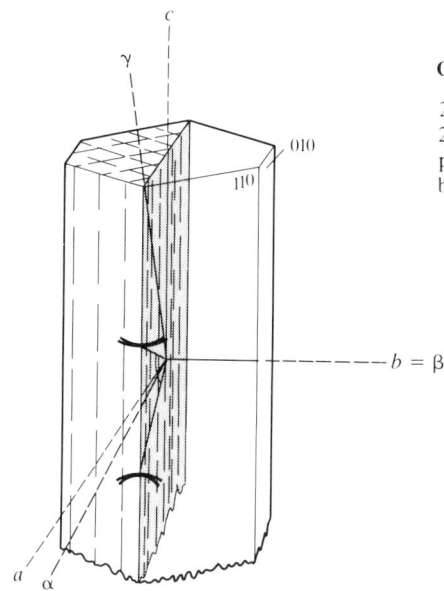

Common hornblende

$2V_\alpha$ variable, usually large, but $2V_\alpha$ for hastingsite is very small: pargasite has γ as acute bisectrix and is +ve

$b = \beta$

$n_\alpha = 1.615–1.705$
$n_\beta = 1.618–1.729$
$n_\gamma = 1.632–1.730$
$\delta = 0.014–0.028$

The large variation in RI is due to compositional differences, particularly the Mg:Fe ratio in the hornblende. Ferric iron and aluminium in the Z sites will also affect both RIs and $2V$

$2V_\alpha = 15°–90°$ −ve (Mg hornblendes are +ve with $2V_\gamma$ almost 90°)
OAP is parallel to (010)
$D = 3.02–3.50 \quad H = 5–6$

COLOUR	Variable, light brown or green but much darker colours for iron-rich varieties.
*PLEOCHROISM	Variable with α pale brown or green, β and γ brown green. Iron-rich varieties have α yellow brown or green, β deep green or blue green, and γ very dark green.
HABIT	Prismatic crystals common, usually elongate.
*CLEAVAGE	Usual amphibole cleavages (see introduction). Partings parallel to $\{100\}$ and $\{001\}$ may also be present.
RELIEF	Moderate to high.
ALTERATION	See introductory section.
BIREFRINGENCE	Moderate: maximum interference colours are low second order blues, but these are frequently masked by the body colour, especially if the hornblende is an iron- or sodium-rich variety.
INTERFERENCE FIGURE *EXTINCTION ANGLE TWINNING	Similar to tremolite–actinolite with extinction angle $\gamma\hat{}\mathrm{cl}$ up to 30° (see introduction).

49

SILICATE MINERALS

*OCCURRENCE Common hornblendes are primary minerals, particularly in intermediate plutonic igneous rocks, although they can occur in other types. In intermediate rocks, the hornblende has a Fe:Mg ratio of about 1:1, whereas hornblendes are more Mg rich in basic rocks and very iron rich in acid rocks (~20:1). Hornblende may occur in some basic rocks (e.g. troctolites etc.) as a corona surrounding olivine crystals, caused by reaction between olivine and plagioclase. Hornblende is stable under a wide range of pressure and temperature (PT) conditions in metamorphism, being an essential constituent of the amphibolite facies. Hornblendes become more alumina rich with increasing metamorphic grade. Pure tschermakite occurs in some high grade metamorphic rocks (often with kyanite) and pure pargasite occurs in metamorphosed dolomites. Secondary amophiboles in igneous rocks are usually tremolites or cummingtonites, but may be hornblendes.

AMPHIBOLE GROUP

Alkali amphiboles
Glaucophane $Na_2(Mg_3Al_2)Si_8O_{22}(OH)_2$
Riebeckite $Na_2(Fe_3^{2+}Fe_2^{3+})Si_8O_{22}(OH)_2$

monoclinic
$0.54:1:0.29$
$\beta = 104°$

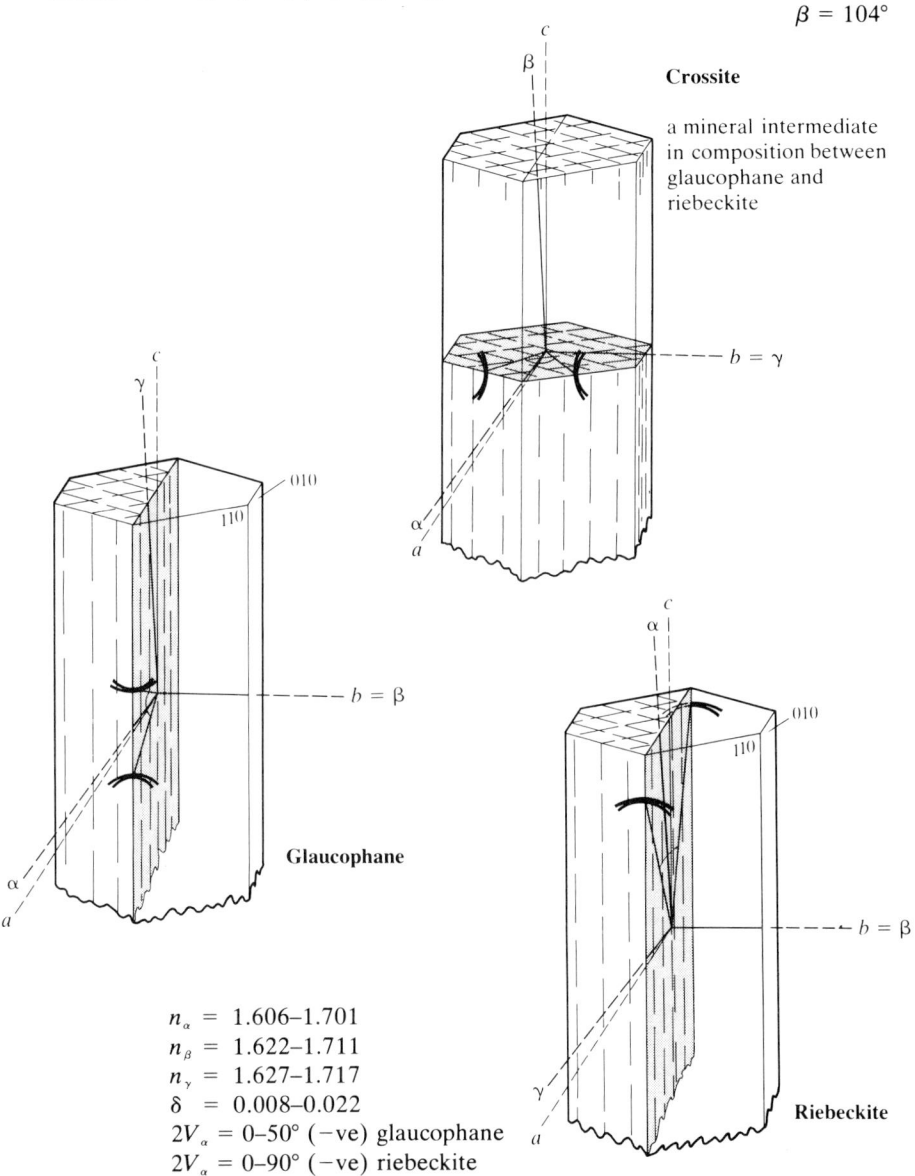

Crossite

a mineral intermediate in composition between glaucophane and riebeckite

Glaucophane

Riebeckite

$n_\alpha = 1.606–1.701$
$n_\beta = 1.622–1.711$
$n_\gamma = 1.627–1.717$
$\delta = 0.008–0.022$
$2V_\alpha = 0–50°\ (-ve)$ glaucophane
$2V_\alpha = 0–90°\ (-ve)$ riebeckite
OAP is parallel to (010) in glaucophane and riebeckite but is perpendicular to (010) in crossite, an intermediate variety
$D = 3.02–3.43$ $H = 6$ (glaucophane), 5 (riebeckite)

SILICATE MINERALS

*COLOUR
: Glaucophane is lavender blue or colourless, whereas riebeckite is dark blue to greenish.

*PLEOCHROISM
: Common in both minerals, with α colourless, β lavender blue, and γ blue in glaucophane, and α blue, β deep blue, and γ yellow green in riebeckite.

HABIT
: Glaucophane occurs usually as tiny blue prismatic crystals whereas riebeckite occurs as either large subhedral prismatic crystals or tiny crystals in the ground mass of some igneous rocks such as alkali microgranites.

CLEAVAGE
: See introduction.

RELIEF
: Moderate to high.

ALTERATION
: Rare in glaucophane; more common in riebeckite, which may alter to a fibrous asbestos (crocidolite). Riebeckite is often found in intimate association with sodic pyroxenes (aegirine), in alkali granites and syenites for example.

BIREFRINGENCE
: Low to moderate; riebeckite interference colours are usually masked by the mineral colour.

INTERFERENCE FIGURE
: The optic axial angles of both minerals may vary considerably in size. In riebeckite the strong colour of the mineral may make the sign very difficult to obtain.

*EXTINCTION ANGLE
: Glaucophane is length slow with a small extinction angle of γ^cleavage (slow^cleavage) of 6–9°. Riebeckite is length fast with an extinction angle of α (fast)^cl = 6–8°. An (010) section in each mineral will give a maximum extinction angle. The variation in extinction angles is caused by the replacement of Al^{3+} by Fe^{3+} in glaucophane and Fe^{2+} in riebeckite.

TWINNING
: Can be simple or repeated on $\{100\}$.

DISTINGUISHING FEATURES
: The lavender blue colour of glaucophane and the fact that it is almost length slow, and the deep blue colour of riebeckite and that it is nearly length fast, are important identification points. Where a mineral has a strong body colour, a mineral edge should be obtained which must be wedge shaped. At the very edge the mineral is so thin that the body colour has a limited effect. Then, using a high powered lens (e.g. ×30), whether the mineral is length fast or length slow can be obtained using a first order red accessory plate.

*OCCURRENCE
: Glaucophane is the essential amphibole in blueschists, which form under high P low T conditions in metamorphosed sediments at destructive plate margins and are commonly found in association with ophiolite suites. Riebeckite occurs in alkali igneous rocks, especially alkali granites where it is associated with aegirine. Fibrous riebeckite (crocidolite, blue asbestos) is formed from the metamorphism at moderate T and P of massive ironstone deposits.

AMPHIBOLE GROUP

Richterite $Na_2Ca(Mg,Fe^{3+},Fe^{2+},Mn)_5Si_8O_{22}(OH,F)_2$ monoclinic

Richterite
Oxyhornblende
Kaersutite
all with α as Bx_a, and large $2V_\alpha$

Katophorite

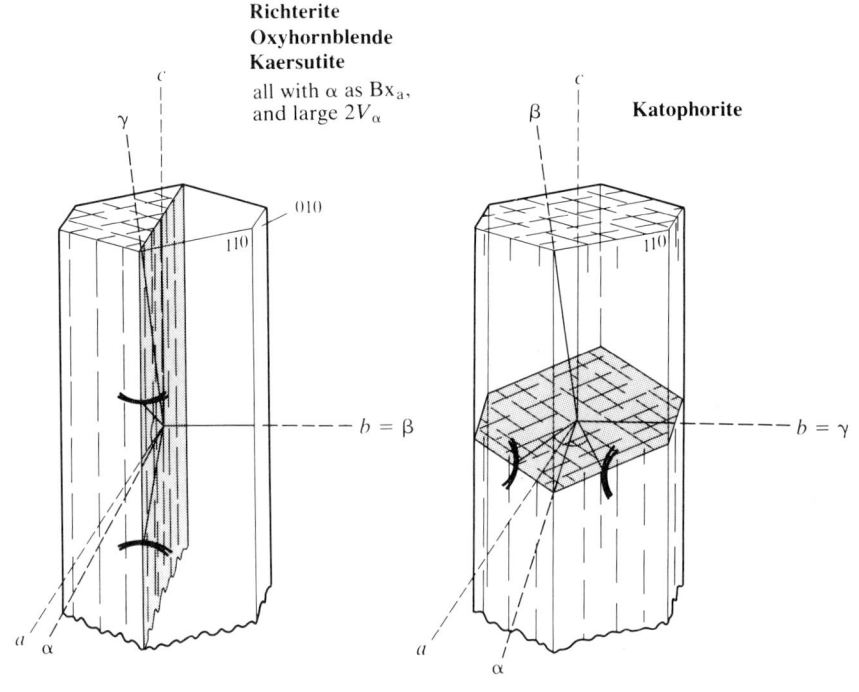

n_α = 1.605–1.685
n_β = 1.618–1.700
n_γ = 1.627–1.712
δ = 0.022–0.027
$2V_\alpha$ = 66°–90° −ve
OAP is parallel to (010)
D = 2.97–3.45   H = 5½

COLOUR Colourless, pale yellow.
*PLEOCHROISM Weak, in pale colours, yellow, orange and blue tints. β is usually darker in colour than α and γ, which are very pale.
HABIT See introduction.
CLEAVAGE Normal, see introduction.
RELIEF Moderate to high.
BIREFRINGENCE Moderate.
INTERFERENCE FIGURE Large 2V on (100) face, but an isotropic section perpendicular to a single optic axis should be obtained and the sign and size of 2V determined from it.
EXTINCTION ANGLE Larger than normal with γ^cleavage 15 to 40° measured in an (010) prism section.

53

SILICATE MINERALS

TWINNING   Simple or repeated on $\{100\}$.
*OCCURRENCE   Rare: formed in metamorphic skarns and in thermally metamorphosed limestones.

The following monoclinic amphiboles are also brown in colour:

Katophorite $Na_2Ca(Mg,Fe)_4Fe^{3+}(Si_7Al)O_{22}(OH)_2$
Oxyhornblende $NaCa_2(Mg,Fe,Fe^{3+},Ti,Al)_5(Si_6Al_2)O_{22}(O,OH)_2$
(basaltic hornblende)
Kaersutite $(Na,K)Ca_2(Mg,Fe)_4Ti(Si_6Al_2)O_{22}(OH)_2$

COLOUR   Oxyhornblende and kaersutite are dark brown in colour.
PLEOCHROISM   Oxyhornblende, α yellow, β and γ dark brown. Kaersutite, δ yellowish, β reddish brown, γ dark brownish. Katophorite is strongly coloured in yellows, browns or greens, with α yellow or pale brown, β greenish brown or dark brown, and γ greenish brown, red brown or purplish brown. In iron-rich varieties β and γ become more greenish and γ may be black.
INTERFERENCE FIGURE   All minerals are negative with $2V_\alpha$ of:

0–50°   katophorite
60–80° $\begin{cases} \text{oxyhornblende} \\ \text{kaersutite} \end{cases}$

EXTINCTION ANGLE   Extinction angles measured on an (010) section vary with composition, as follows:

$\beta\hat{\ }$cleavage 20° to 54°   katophorite
$\gamma\hat{\ }$cleavage 0 to 19° $\begin{cases} \text{oxyhornblende} \\ \text{kaersutite} \end{cases}$

SUMMARY OF PROPERTIES   Katophorite is very strongly coloured and pleochroic in yellows, browns and greens, and with $2V_\alpha$ variable (0–50°) and a large extinction angle $\beta\hat{\ }$cl = 20 to 54° on an (010) section. Note that the OAP is perpendicular to (010).

Oxyhornblende is pleochroic in yellows and dark browns, and with $2V_\alpha$ large and with a small angle $\gamma\hat{\ }$cl = 0 to 19' on an (010) section.

Kaersutite is pleochroic in yellows and reddish browns, and with $2V_\alpha$ large. Extinction angles are small with $\gamma\hat{\ }$cl 0 to 19° on an (010) section.

OCCURRENCE   Katophorite occurs in dark coloured alkali intrusives in association with nepheline, aegirine and arfvedsonite. Kaersutite occurs in alkaline volcanic rocks, and as phenocrysts in trachytes and other K-rich extrusives; and it may be present in some monzonites.

Oxyhornblende occurs mainly as phenocrysts in intermediate volcanic or hypabyssal rocks such as andesites, trachytes and so on.

AMPHIBOLE GROUP

Note that the mineral barkevikite is no longer recognised as a distinct mineral and the name has been formally abandoned. Barkevikite was a name used to describe an iron-rich pargasitic hornblende, and was never chemically defined (Leake 1978).

Eckermannite–arfvedsonite $Na_2Na(Mg,Fe^{2+})_4AlSi_8O_{22}(OH,F)_2$ monoclinic

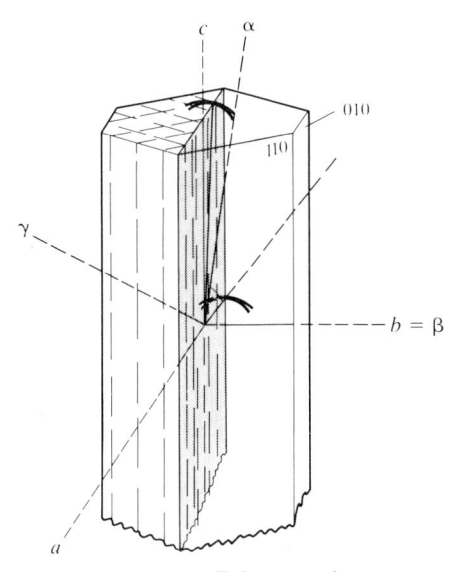

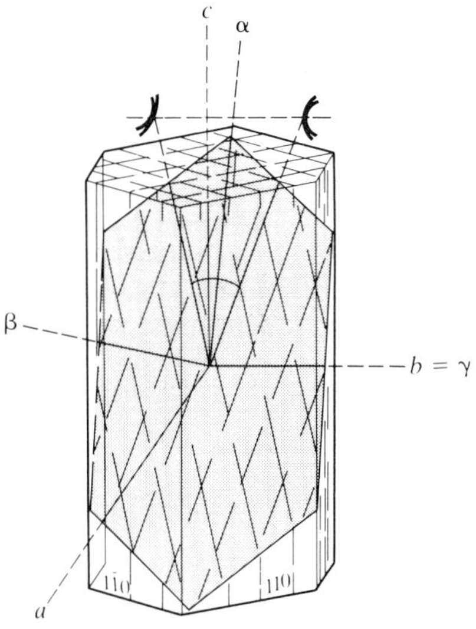

Eckermannite
n_α = 1.612–1.638
n_β = 1.625–1.652
n_γ = 1.630–1.654
δ = 0.009–0.020
$2V_\alpha$ = 80°–15° −ve
OAP is parallel to (010)
D = 3.00–3.16   H = 5½

Arfvedsonite
n_α = 1.674–1.700
n_β = 1.679–1.709
n_γ = 1.686–1.710
δ = 0.005–0.012
$2V_\alpha$ = variable, probably −ve
OAP is perpendicular to (010)
D = 3.30–3.50   H = 5½

COLOUR Eckermannite is pale green and arfvedsonite has strong shades of green.
*PLEOCHROISM Eckermannite is pleochroic with α blue green, β light green, and γ pale yellowish green; and arfvedsonite has α greenish blue to indigo, β lavender blue to brownish yellow, and γ greenish yellow to blue grey.
HABIT Both minerals occur as large subhedral prisms, often corroded along the edges and frequently poikilitically enclosing earlier crystallising ferromagnesian minerals.
CLEAVAGE Normal (see introduction).
RELIEF Moderate (Eck) to high (Arfv).
ALTERATION Common, to chloritic minerals.
*BIREFRINGENCE Low in both minerals but interference colours are frequently masked by mineral colours, especially in arfvedsonite.

SILICATE MINERALS

*INTERFERENCE FIGURE	The colour of the minerals and their strong dispersion make interference figures difficult to obtain, and these are usually indistinct with optic signs and size of $2V$ impossible to judge.
*EXTINCTION	Oblique with both minerals having variable extinction angles; $\alpha\char`\^$cleavage varies from $0°$ to $50°$ but this is also difficult to obtain.
TWINNING	Simple or repeated on $\{100\}$.
OCCURRENCE	Both minerals occur as constituents in alkali plutonic rocks (soda-rich rocks), such as nepheline– and quartz–syenites, where they occur in association with aegirine or aegirine–augite and apatite. The minerals are late crystallisation products.
Aenigmatite	Aenigmatite ($Na_2Fe_5^{2+}TiSi_6O_{20}$) is a mineral closely resembling the alkali amphiboles. It has very high relief (~ 1.8) and a small positive $2V$. Aenigmatite is pleochroic with α red brown, β brown, and γ dark brown. It is similar to the dark brown amphiboles but has higher RIs. Aenigmatite often occurs as small phenocrysts in alkaline volcanic rocks such as phonolites.

Beryl   Cyclosilicate

Beryl $Be_3Al_2Si_6O_{18}$ hexagonal
                                                        c/a 0.9956

$n_o = 1.560–1.602$
$n_e = 1.557–1.599$
$\delta = 0.003–0.009$
Uniaxial + ve (a prism section is length fast)
$D = 2.66–2.92$ $H = 7½–8$

COLOUR	Colourless, pale yellow or pale green.
PLEOCHROISM	Weakly pleochroic in pale greens if section is thick.
*HABIT	Hexagonal prism with large basal face.
CLEAVAGE	Imperfect basal $\{0001\}$.
RELIEF	Low to moderate.
ALTERATION	Beryl easily undergoes hydrothermal alteration to clay minerals, as follows, the reaction releasing quartz and phenakite:

$$2Be_3Al_2Si_6O_{18} + 4H_2O \rightarrow Al_4Si_4O_{10}(OH)_8 + 5SiO_2 + 2Be_2SiO_4$$
$$\text{kaolin} \hspace{4em} \text{phenakite}$$

BIREFRINGENCE	Low first order greys.
TWINNING	Rare
*OCCURRENCE	Beryl occurs in vugs in granites and particularly in pegmatites, often associated with cassiterite. The precious stone variety, aquamarine, occurs in similar locations, but emerald is usually found in metamorphic biotite schists.

Chlorite

Chlorite $(Mg,Al,Fe)_{12}(Si,Al)_8O_{20}(OH)_{16}$

Phyllosilicate

monoclinic
$0.57:1:1.31, \beta = 97°$

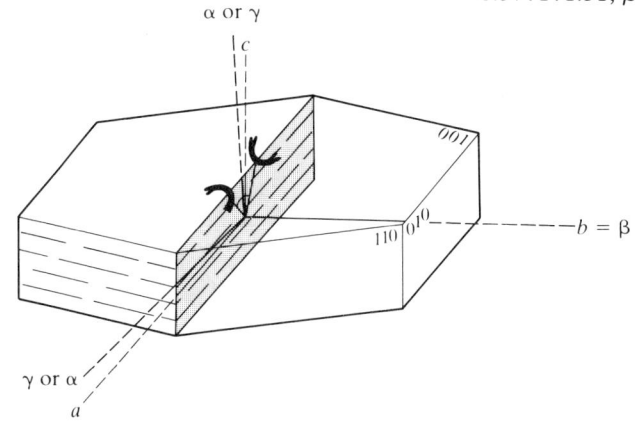

$n_\alpha = 1.57–1.66$
$n_\beta = 1.57–1.67$
$n_\gamma = 1.57–1.67$
$\delta = 0.0–0.01$
$2V = 20°–60°$ +ve or −ve
OAP is parallel to (010)
$D = 2.6–3.3 \quad H = 2–3$

*COLOUR	Colourless or green.
PLEOCHROISM	Green varieties have α pale green to colourless, β and γ darker green.
HABIT	Tabular crystals with a pseudo-hexagonal shape.
*CLEAVAGE	Perfect $\{001\}$ basal cleavage.
RELIEF	Low to moderate.
ALTERATION	Oxidation of iron in chlorite may occur (the sign changes from +ve to −ve).
*BIREFRINGENCE	Very weak, usually with anomalous deep *Berlin blue* colour.
INTERFERENCE FIGURE	Biaxial Bx_a figure on basal section with small $2V$. Usually positive but some varieties – chamosite in particular – are optically negative. Interference figures are rarely obtained.
EXTINCTION	Straight to cleavage but can be oblique with small angle γ or $\alpha\char`\^cl$ (fast or slow to cleavage); very small angle ($< 5°$) on (010) section.
TWINNING	As in micas: rare.
OCCURRENCE	Chlorite is a widely distributed primary mineral in low grade regional metamorphic rocks (greenschists), eventually changing to biotite with increasing grade; muscovite is also involved in the reaction. The initial material is usually argillaceous sediments, but basic igneous rocks and tuffs will give chlorite during regional metamorphism. In some alkali-rich rocks, chlorite will break down with increasing P and T and help to

SILICATE MINERALS

form amphibole and plagioclase. In igneous rocks chlorite is usually a secondary mineral, forming from the hydrothermal alteration of pyroxenes, amphiboles and biotites. Chlorite may be found infilling amygdales in lavas with other minerals, and may occur as a primary mineral in some low temperature veins.

Chlorites are common in argillaceous rocks where they frequently occur with clay minerals, particularly illite, kaolin and mixed-layer clays.

Chloritoid — Nesosilicate

Chloritoid (ottrelite) $(Fe,Mg)_2(Al,Fe^{3+})Al_3O_2[SiO_4]_2(OH)_4$ monoclinic
$1.725:1:3.314, \beta = 101°30'$

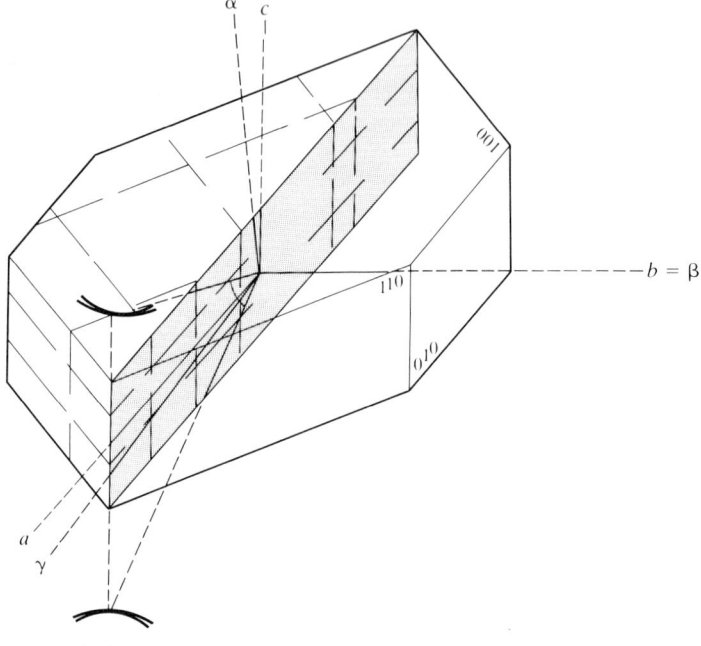

$n_\alpha = 1.713–1.730$
$n_\beta = 1.719–1.734$
$n_\gamma = 1.723–1.740$
$\delta = 0.010$
$2V_\gamma = 45°–68°$ +ve (normal range). $2V$ can be highly variable with $2V_\gamma$ 36°–90° +ve and $2V_\alpha$ 90°–55° −ve
OAP is parallel to (010)
$D = 3.51–3.80$ $H = 6½$

COLOUR — Colourless, green, blue green.
*PLEOCHROISM — Common with α pale green, β blue and γ colourless to pale yellow.
HABIT — Closely resembles mica minerals, occurring as pseudo-hexagonal tabular crystals.

*CLEAVAGE	$\{001\}$ perfect. Another poor possibly prismatic fracture may be present which distinguishes chloritoid from micaceous minerals.
RELIEF	High.
ALTERATION	Chloritoid may alter to muscovite and chlorite, but this is not common.
BIREFRINGENCE	Low but masked by greenish colour of mineral, often anomalous blue colours are seen.
INTERFERENCE FIGURE	A bluish green coloured (100) section of chloritoid will give a Bx_a figure with a moderate $2V$ and positive sign.
EXTINCTION	Straight to perfect on $\{001\}$ cleavage.
*OTHERS	Zoning occasionally appears as a peculiar hourglass shape seen on prismatic sections.
OCCURRENCE	Chloritoid occurs in regionally metamorphosed pelitic rocks with a high $Fe^{3+}:Fe^{2+}$ ratio, at low grades of metamorphism. Chloritoid develops about the same time as biotite, changing to staurolite at higher grades. Chloritoid can occur in non-stress environments where it usually shows triclinic crystal form, particularly in quartz carbonate veins and in altered lava flows.

Clay minerals — Phyllosilicates

The clay minerals are extremely important in weathering processes. Many primary igneous minerals produce clay minerals as a final weathering product. Feldspars particularly give rise to clay minerals; plagioclase feldspar reacts with water to give montmorillonite, and orthoclase feldspar in a similar way produces illite. If excess water is present both montmorillonite and illite will eventually change to kaolin, which is always the final product.

Kaolin (kandite) $Al_4Si_4O_{10}(OH)_8$ triclinic
0.576 : 1 : 0.830
$\alpha = 91°48', \beta = 104°30', \gamma = 90°$

$n_\alpha = 1.553-1.565$
$n_\beta = 1.56-1.57$
$n_\gamma = 1.56-1.57$
$\delta = 0.006$
$2V_\alpha = 24°-50°$ −ve
OAP perpendicular to (010)
$D = 2.61-2.68$ $H = 2-2\frac{1}{2}$

COLOUR	Colourless.
*HABIT	Similar to mica group, but crystals are extremely tiny.
RELIEF	Low.
CLEAVAGE	Perfect basal – similar to micas.
BIREFRINGENCE	Low, greys of first order.
INTERFERENCE FIGURE	Size of individual crystals is such that interference figures can rarely be obtained.
EXTINCTION	Straight but occasional slight extinction angle on (010) face.

SILICATE MINERALS

Grain size of all clay minerals is extremely small and optical determinations are generally useless. Serious studies and precise identification of clay minerals are carried out either by X-ray diffraction techniques (XRD) or by using a scanning electron microscope (SEM) or electron microprobe.

The occurrence of all clay minerals will be discussed together after montmorillonite.

Illite $K_{1-1.5}Al_4[Si_{7-6.5}Al_{1-1.5}O_{20}](OH)_4$ monoclinic

n_α = 1.54–1.57
n_β = 1.57–1.61
n_γ = 1.57–1.61
δ = 0.03
$2V$ = small ($< 10°$) −ve
OAP approx. parallel to (010)
D = 2.6–2.9 H = 1–2

Properties similar to kaolin with the exception of birefringence, which is much stronger with second order colours.

Montmorillonite group (smectites) $(½Ca,Na)_{0.7}(Al,Mg,Fe)_4(Si,Al)_8O_{20}(OH)_4 \cdot nH_2O$ monoclinic

n_α = 1.48–1.61
n_β = 1.50–1.64
n_γ = 1.50–1.64
δ = 0.01–0.04
$2V_\alpha$ = small −ve
OAP is parallel to (010)

Properties similar to those of kaolin and illite.

OCCURRENCE Kaolin is the most common of the clay minerals and forms by hydrothermal alteration or weathering of feldspars, feldspathoids and other silicates. Kaolin, therefore, usually forms from the alteration of acid igneous rocks (granites etc.), with non-alkaline conditions being required.

Illite is the common clay mineral in clays and mudstones, and it is formed by weathering of feldspars or by alteration of other clay minerals during sediment formation. Illite formation is favoured by alkaline conditions and high Al and K activities.

Montmorillonite and smectites are principal constituents of bentonite clays, formed from alteration of pyroclastic ash deposits (tuff etc.). Montmorillonite (particularly Fuller's earth) is formed by alteration of basic igneous rocks in areas of poor drainage when Mg is not removed. An alkaline environment is preferred, with low K and higher Ca. Vermiculite is another clay mineral derived from biotite alteration, and with properties closely related to the smectite group.

Cordierite   Cyclosilicate

Cordierite $Al_3(Mg,Fe)_2Si_5AlO_{18}$ orthorhombic
pseudo-hexagonal
$0.568 : 1 : 0.549$

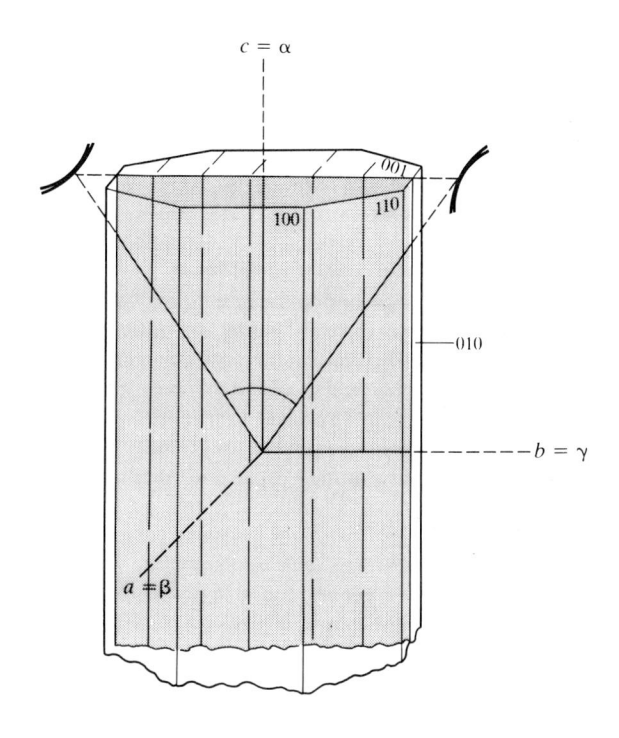

n_α = 1.522–1.558
n_β = 1.524–1.574
n_γ = 1.527–1.578
δ = 0.005–0.020
$2V_\gamma$ = 65°–90° +ve
$2V_\alpha$ = 90°–76° −ve
OAP is parallel to (100)
D = 2.53–2.78 H = 7

COLOUR   Colourless, but occasionally blue. The precious variety of cordierite (water sapphire) is pale violet and pleochroic.

PLEOCHROISM   Iron-rich varieties show α colourless, β and γ violet, whereas magnesium-rich varieties only show pleochroism in thick sections with α pale yellow green, β and γ pale blue.

*HABIT   In regionally metamorphosed rocks, cordierite occurs as large 'spongy' crystals containing inclusions of muscovite, biotite and quartz. Almost

all the crystal shows alteration, and no good crystal face edges occur. In thermal aureoles, cordierite occurs in inner zones, and the fresh cordierite crystals which develop often show good hexagonal crystal form. Cordierite may occur in some igneous rocks, where it shows subhedral to euhedral crystal form.

CLEAVAGE {010} good, with a poor {001} basal cleavage sometimes developing.

*RELIEF Low, similar to quartz; usually higher than 1.54.

*ALTERATION Cordierite shows alteration at edges and along cracks to pinite (a mixture of fine muscovite and chlorite or serpentine), which usually appears as a pale yellowish green mineral in thin section. A yellow pleochroic halo may sometimes appear in a cordierite crystal, surrounding an inclusion of zircon or monazite, similar to those found in biotite crystals. Such haloes are caused by the elements of the radioactive series U–Ra and Th–Ac.

BIREFRINGENCE Low, similar to quartz or feldspar.

INTERFERENCE FIGURE $2V$ is very large, and so the best figure would be obtained by examining a near isotropic section giving an optic axis figure. Such a figure is approximately found in the position of face (011) or (0T1).

*TWINNING Extremely common in all crystals except those in regional metamorphic rocks, where, in any case, alteration masks any twinning. Fresh, clear crystals in thermal aureoles, and some 'partial melt' igneous rocks, show two kinds of twinning – cyclic and lamellar. Cyclic twinning on {110} or {130} produces a pseudo-trigonal or pseudo-hexagonal pattern. In some instances twinning on these planes produces lamellar twinning similar to twinning seen in plagioclase feldspars.

Cordierite is a mineral found in pelitic rocks which have been subjected to metamorphism at low pressure. Cordierite occurs in the inner (high temperature) zone of thermal aureoles, and in regional metamorphic conditions of high heat flow and low pressure, such as Buchan-type metamorphism, where the sequence of index minerals produced under regional conditions is, progressively: biotite–andalusite–cordierite–sillimanite. In these rocks, cordierite occurs in high grade gneisses, either under abnormal PT conditions, or where a thermal metamorphic episode follows regional metamorphism, and pre-existing minerals such as kyanite and biotite become unstable, reacting to give cordierite and muscovite.

*OCCURRENCE Cordierite may occur in some igneous rocks, especially cordierite–norites. Originally they were considered to represent the crystallised products of basic magma contaminated by the assimilation of argillaceous material. However, it has recently been suggested that such rocks represent partial melt products in which high temperature liquids have formed from the pelitic (or argillaceous) rocks owing to extremely high temperatures being developed from nearby emplaced basic intrusions, these crystallising to give cordierite–norites. In these partial melt rocks there is no magmatic component. Cordierite has been known to occur in some granites and granite pegmatites as a primary mineral.

Epidote group

Sorosilicates

α-zoisite has *ac* as the optic axial plane.
β-zoisite has *ab* as the optic axial plane.
All monoclinic epidotes have *ac* as the optic axial plane.

Minerals in the epidote group belong to both the orthorhombic and the monoclinic systems. Conventionally the mineral belonging to the higher symmetry system (orthorhombic) is described first, and therefore the descriptions begin with zoisite and go on to the two important monoclinic varieties, clinozoisite and epidote. Two other varieties – piemontite (or piedmontite), a manganese-bearing epidote, and allanite (or orthite), a cerium-bearing type – are not described in detail here as they are relatively rare, allanite occasionally being found as an accessory mineral in some syenites and granites.

Zoisite $Ca_2Al_3(OH)Si_3O_{12}$

orthorhombic
0.622 : 1 : 0.347

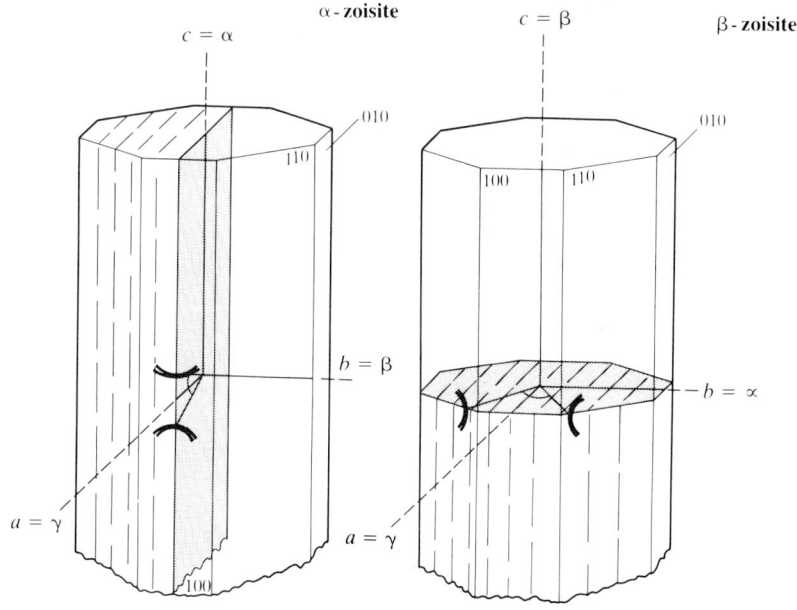

$n_\alpha = 1.696$
$n_\beta = 1.696$
$n_\gamma = 1.702$ } The RIs may vary depending on the amount of trace elements (Fe^{3+} etc.) in the structure

$\delta = 0.006$
$2V_\gamma = 0-60°$ +ve

OAP is either parallel to (010) in α-zoisite, or parallel to (001) in β-zoisite

$D = 3.15-3.36$ $H = 6$

SILICATE MINERALS

COLOUR	Colourless; a pink variety (thulite) may occur in Mn-rich environments.
HABIT	Usually found in clusters of elongate prismatic crystals, with rectangular cross sections.
CLEAVAGE	Perfect $\{100\}$ prismatic cleavage, poor $\{001\}$ cleavage sometimes present.
*RELIEF	High.
ALTERATION	Since zoisite forms under conditions of low P and T, it remains stable and is not subject to further reactions. Zoisite may form from the breakdown of calcium plagioclase from basic igneous rocks which have suffered hydrothermal alteration (a process called saussuritisation).
*BIREFRINGENCE	Very low (varies between 0.004 and 0.008). α-zoisite shows low first order colours (greys, whites) but β-zoisite shows anomalous interference colours of a deep Berlin blue.
EXTINCTION	Straight on prism edge or $\{100\}$ cleavage.
*INTERFERENCE FIGURE	A (100) section (001) gives a biaxial positive figure with a moderate $2V$.
OTHER FEATURES	Since β-zoisite differs from α-zoisite in containing up to 5% Fe_2O_3, an intermediate variety between α- and β-zoisite may occur, which is distinguished by possessing a very small $2V$ ($\approx 0°$).
OCCURRENCE	Zoisite is found in basic igneous rocks which have been hydrothermally altered, where it develops from calcic plagioclase. It also occurs in medium grade metamorphosed schists in association with sodic plagioclase, amphibole, biotite and garnet. It may occur in some metamorphosed impure limestones.

Clinozoisite $Ca_2Al_3(OH)Si_3O_{12}$ monoclinic
                                                       $1.583:1:1.814$, $\beta = 115°30'$

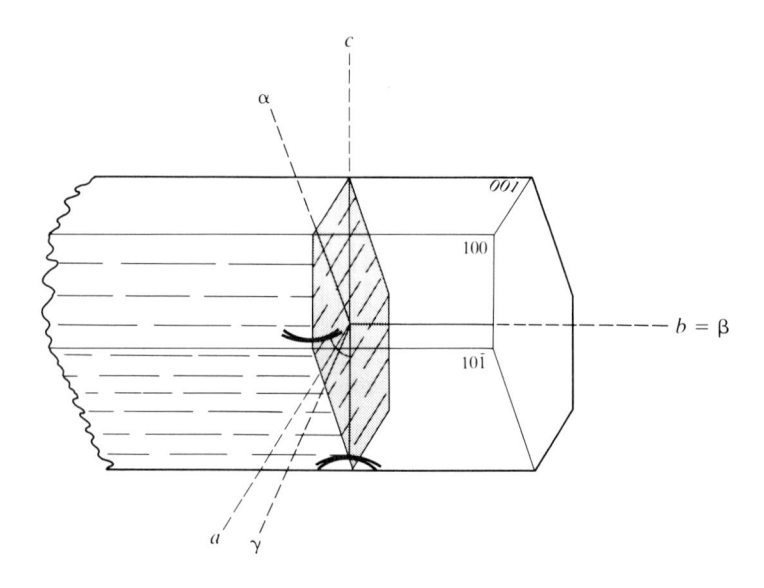

EPIDOTE GROUP

$n_\alpha = 1.710$
$n_\beta = 1.715$
$n_\gamma = 1.719$
RIs are based on a clinozoisite with about 1 per cent Fe_2O_3 present. These will increase with increasing Fe_2O_3 content or decrease if clinozoisite is iron free

$\delta = 0.005–0.015$ (variable with Fe_2O_3 content)

$2V_\gamma$ = variable, usually 14°–90° +ve

OAP is parallel to (010)

$D = 3.12–3.38 \quad H = 6½$

COLOUR	Colourless.
HABIT	Found in columnar aggregates of crystals, which are usually quite small.
CLEAVAGE	Perfect {001} cleavage, appearing as a prismatic cleavage in sections, since mineral is elongate parallel to b axis.
*RELIEF	High.
ALTERATION	None.
*BIREFRINGENCE	Very low with anomalous first order interference colours (deep blue, greenish yellow: no first order white).
EXTINCTION	Oblique, variable extinction angles depending on mineral composition, but most elongate prismatic sections have straight extinction on cleavage.
*INTERFERENCE FIGURE	A (100) section will give a biaxial positive figure, but since $2V$ is large an isotropic section should be selected and a single isogyre examined for sign and size.
OCCURRENCE	Clinozoisite occurs primarily in regionally metamorphosed low grade rocks forming from micaceous minerals. Its other occurrences are similar to those of zoisite.

SILICATE MINERALS

Epidote (pistacite) $Ca_2Fe^{3+}Al_2(OH)Si_3O_{12}$ monoclinic
$1.591:1:1.812, \beta = 115°24'$

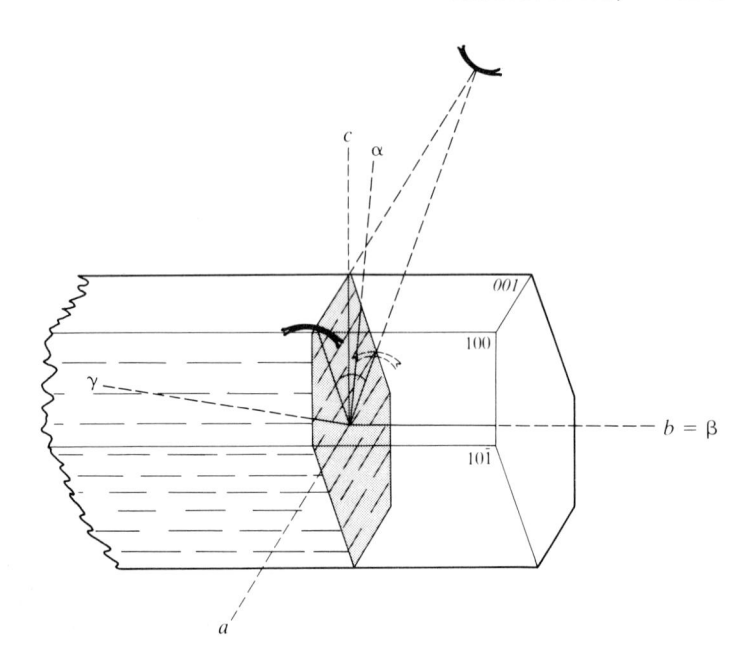

$n_\alpha = 1.715$
$n_\beta = 1.725$ } RIs increase with increasing Fe^{3+} content
$n_\gamma = 1.734$
$\delta = 0.019–0.049$ (variable with composition)
$2V_\alpha =$ variable $64°–90°$ $-ve$
OAP is parallel to (010)
$D = 3.38–3.49$ $H = 6$

COLOUR — Colourless to pale yellowish green.
*PLEOCHROISM — Slightly pleochroic with α colourless to pale yellowish green, β greenish and γ yellowish.
HABIT — Found in aggregates of elongate prismatic crystals with pseudo-hexagonal cross sections.
CLEAVAGE — {001} perfect (similar to clinozoisite).
*RELIEF — High.
ALTERATION — None.
*BIREFRINGENCE — Moderate to high, showing low second order to upper third order colours. Some sections may show low anomalous interference colours similar to clinozoisite.
EXTINCTION — Oblique to cleavage in pseudo-hexagonal sections (see figure), otherwise straight on cleavage in prismatic section.

INTERFERENCE FIGURE	A section perpendicular to the c axis will give a biaxial negative figure with a large $2V$; thus an optic axis figure is preferable when determining the sign.
OTHER FEATURES	Lamellar twinning may sometimes be present.
*OCCURRENCE	Epidote is an important mineral in low grade regional metamorphic rocks, where it marks the beginning of the epidote–amphibolite facies, forming from the breakdown of chlorite. Epidote also forms from saussuritisation of plagioclase feldspar and from the breakdown of amphiboles in basic igneous rocks, these changes being due to late stage hydrothermal alteration. In highly amphibolitised basic igneous rocks, clusters of epidote crystals are commonly seen associated with plagioclase feldspar and often within amphibole minerals.

Feldspar group   Tektosilicates

Introduction

The feldspars are the most important minerals in rocks. They occur in igneous, metamorphic and sedimentary rocks, and their range of composition has led to their use as a means of classifying igneous rocks. Feldspars are absent only from certain ultramafic and ultra-alkaline igneous rock types and carbonatites. Feldspars occur in almost all metamorphic rocks, being absent only from some low grade pelitic types, pure marbles and pure quartzites. In sedimentary rocks feldspars are common constituents of many arenaceous rocks, but are less common in shales (clay rocks) and mudstones. Feldspars are difficult to detect (although XRD investigation would reveal their presence) because of the minute grain size (< 0.002 mm) of these argillaceous rocks.

Although the most ubiquitous of minerals, feldspars have a restricted range of composition. There are two main types of feldspars:

(a) Alkali feldspars, which range between the end members orthoclase $KAlSi_3O_8$ and albite $NaAlSi_3O_8$.
(b) Plagioclase feldspars, which range between the end members albite $NaAlSi_3O_8$ and anorthite $CaAl_2Si_2O_8$.

From this it is obvious that albite is common to both feldspar types, and the most usual way of depicting the complete feldspar group is in a ternary (or triangular) diagram with orthoclase, albite and anorthite representing the composition of each apex (Fig. 2.7). This reveals that the alkali feldspars can contain up to 10% anorthite molecule in their structure, and similarly plagioclase feldspars can contain up to 10% orthoclase molecule in their structure.

The optical properties and structure of the feldspars depend upon their temperature of crystallisation and their cooling history. Thus for

SILICATE MINERALS

Figure 2.7 Feldspar composition diagram.

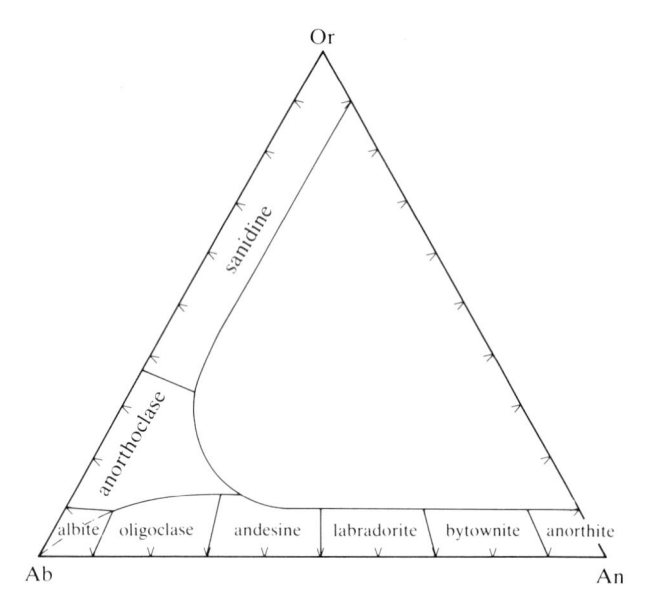

example in quickly cooled extrusive igneous rocks, the potassium feldspar that forms has a tabular crystal habit and is called sanidine, with optical properties peculiar to that mineral. However, when plutonic igneous rocks crystallise, their cooling rate is much slower, which controls the kinetics of order–disorder in the feldspar structure, leading to the formation of orthoclase, with a prismatic habit and slightly different optical properties from those of sanidine.

Much work has been carried out on the feldspar minerals in recent years, and Smith (1974) summarised the differences between the different feldspar types. Figures 2.8, 2.9 and 2.10 illustrate the composition and nomenclature of the feldspars at various temperatures. The term 'temperatures' refers to the temperatures at which effective structural re-equilibration ceases, and the mineral structure will not be subject to any change thereafter. It is a time–temperature–kinetic relationship. The feldspars are divided into three groups – high (representing extrusive rocks), intermediate (representing hypabyssal rocks and small intrusions) and low (deep seated plutonic rocks and metamorphic rocks).

Figure 2.8 shows feldspars quenched from high temperature. An arbitrary boundary from albite (Ab) with anorthite (An) equal to orthoclase (Or) defines the alkali feldspar and plagioclase feldspar fields. The plagioclase feldspars are divided into six divisions at 10, 30, 50, 70 and 90 mol per cent An, and the feldspars in these divisions are named on the diagram as (high) albite 0–10% An, oligoclase 10–30% An, andesine, labradorite, bytownite and anorthite 90–100% An. In the alkali feldspars, the boundary at 40% Or between high albite and high

Figure 2.8 High temperature feldspars.

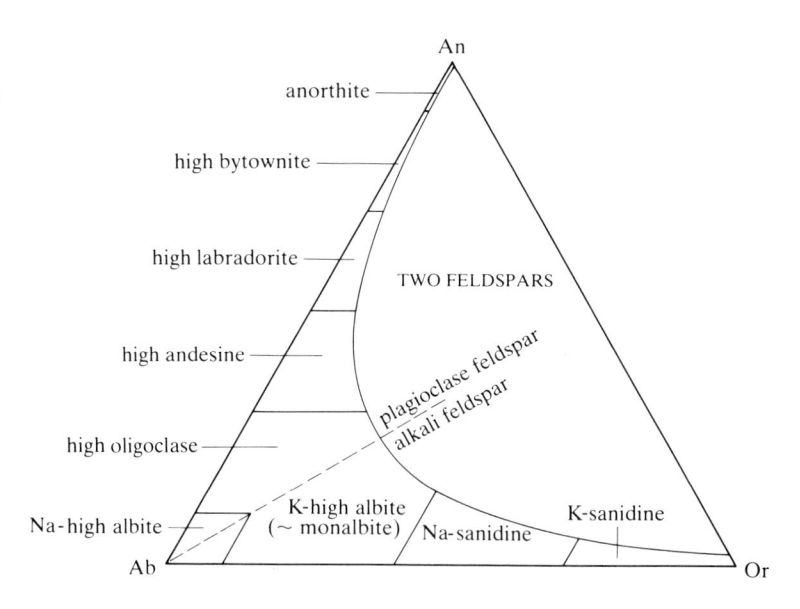

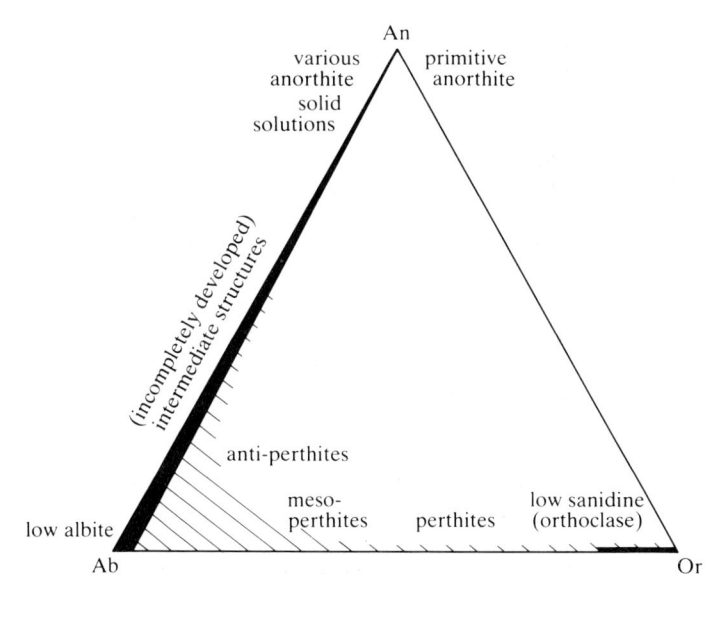

Figure 2.9 Feldspars formed after fast cooling.

SILICATE MINERALS

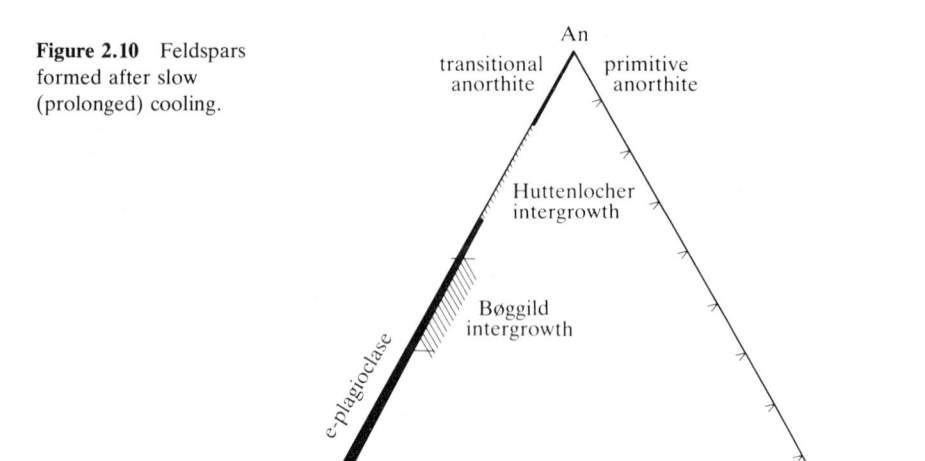

Figure 2.10 Feldspars formed after slow (prolonged) cooling.

[Ternary diagram with An at top, low albite at bottom left, Or at bottom right. Labels include: transitional anorthite, primitive anorthite, Huttenlocher intergrowth, Bøggild intergrowth, e-plagioclase, peristerite intergrowth, low microcline.]

Na-sanidine represents a structural change from triclinic to monoclinic. The boundary at 70% Or separating Na-sanidine from K-sanidine is an arbitrary one.

In Figure 2.9 the feldspars formed after short cooling histories (fast rate of cooling) show unmixing with perthite development. Homogeneous feldspars (similar to those found in Fig. 2.8) occupy small areas in Figure 2.9, with K-feldspar (or orthoclase) restricted to a small field. Most alkali feldspars are perthites, consisting dominantly of an akali feldspar host with an exsolved plagioclase feldspar phase resulting from segregation from an original higher temperature homogeneous feldspar. When plagioclase feldspar is dominant the unmixed feldspar is called an antiperthite. Some perthites consist of roughly equal amounts of intergrown alkali feldspar and plagioclase, and these are called mesoperthites (Fig. 2.9). Submicroscopic cryptoperthites may occur, for example in orthoclase, which tend to give the host mineral optical properties between it (the host) and the exsolved phase.

Figure 2.10 shows the feldspar fields after prolonged cooling histories such as will occur in large plutons or in metamorphic rocks. K-feldspar is restricted to low microcline and homogeneous plagioclases are restricted to nearly pure albite, anorthite An_{85-100} and the compositional range from An_{15} to An_{70}, with all these having no more than about 2 mol% Or in the structure. The non-homogeneous types consist of complex series of intergrowths of which three are important: peristerites (containing equal amounts of alkali feldspar and plagioclase); Bøggild intergrowths (roughly from An_{40} to An_{60}); and Huttenlocher intergrowths, from An_{70} to An_{85}.

The Bøggild intergrowths can contain up to 6% Or and are distinguished by labradorite iridescence (caused by the structure).

Although the individual feldspar types will be described in detail, the general optical properties for feldspars are given below.

COLOUR — Colourless with occasional white or pale brown patches where alteration to clay minerals has occurred.

HABIT — Euhedral crystals – tabular, or prismatic with large basal faces – may occur as phenocrysts in some extrusive rocks, but most feldspars are either subhedral (prismatic) or anhedral in most rocks.

CLEAVAGE — All feldspars possess two cleavages $\{001\}$ and $\{010\}$, intersecting nearly at right angles on a (100) section. Several partings may occur.

RELIEF — Low from just below 1.54 (K-feldspar) to above 1.54 (most plagioclases). Figure 2.14 gives details.

ALTERATION — All feldspars may alter to clay minerals. The individual descriptions give details.

BIREFRINGENCE — Maximum interference colours are first order white in Ca-poor plagioclase, and first order yellow in Ca-rich plagioclase.

INTERFERENCE FIGURES — Variable in sign and size. Usually large, so that a single optic axis figure is often required for examination.

EXTINCTION — Figure 2.12 gives details for alkali feldspar. Plagioclase feldspars show repeated twinning, and the symmetrical extinction angles measured on the twin plane are used to obtain plagioclase composition. Figures 2.16 and 2.17 show these extinction angles, but the relevant section in the plagioclase feldspar descriptions must be consulted for details of this technique.

TWINNING — K-rich alkali feldspars exhibit simple twinning, but the plagioclase feldspars show polysynthetic twinning or repeated twinning or complex multiple twins.

ZONING — Common in most plagioclase feldspars, particularly in phenocrysts in extrusive rocks.

PERTHITES — Feldspars frequently show effects of unmixing or exsolution resulting in intergrowths – plagioclase feldspars within alkali feldspar host and *vice versa*.

SILICATE MINERALS

Alkali feldspars

There is a continuous series from K-feldspar ($KAlSi_3O_8$) to albite ($NaAlSi_3O_8$) for high and low alkali feldspars.

Sanidine–high albite series
Ab_0–Ab_{63} sanidine
Ab_{63}–Ab_{90} anorthoclase
Ab_{90}–Ab_{100} high albite

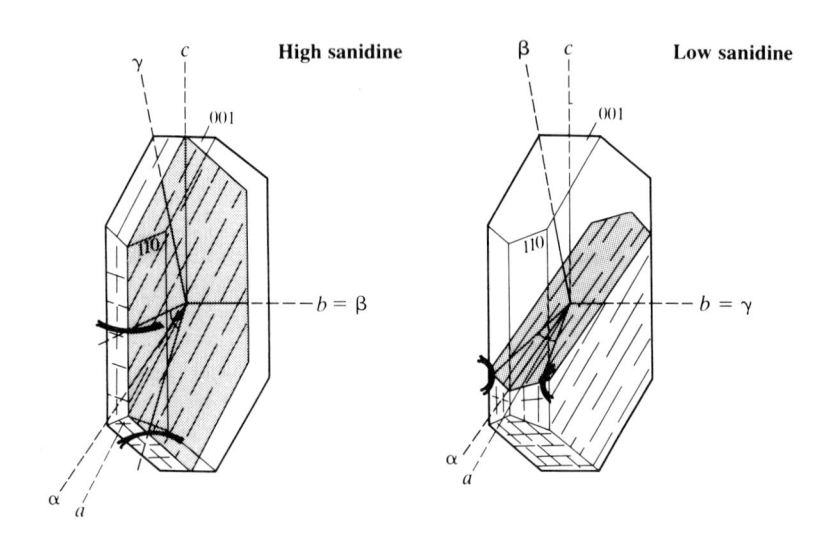

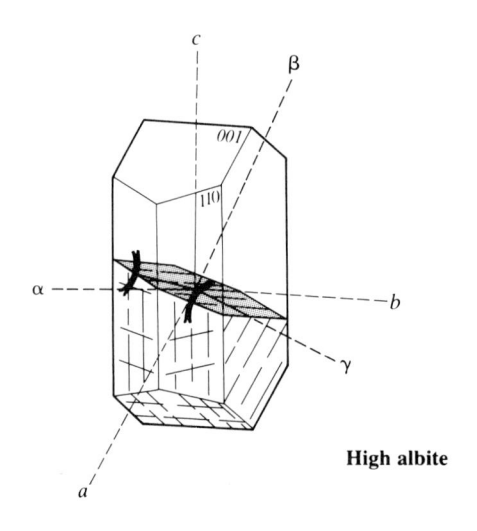

High sanidine

Low sanidine

High albite

FELDSPAR GROUP

Orthoclase–low albite series
- Or_{100-85} orthoclase
- Or_{85-20} orthoclase cryptoperthites
- Or_{20-0} low albite

Microcline–low albite series
- Or_{100-92} microcline
- Or_{92-20} microcline cryptoperthites
- Or_{20-0} low albite

Orthoclase

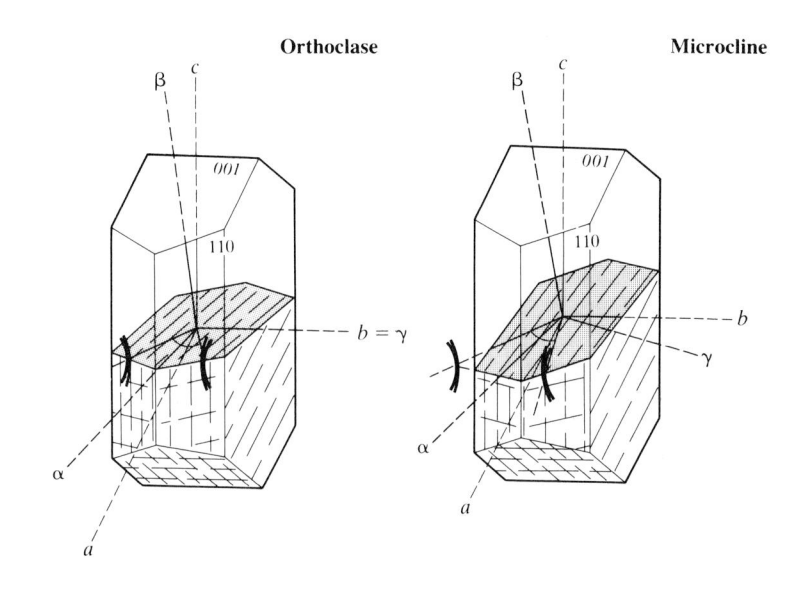

Microcline

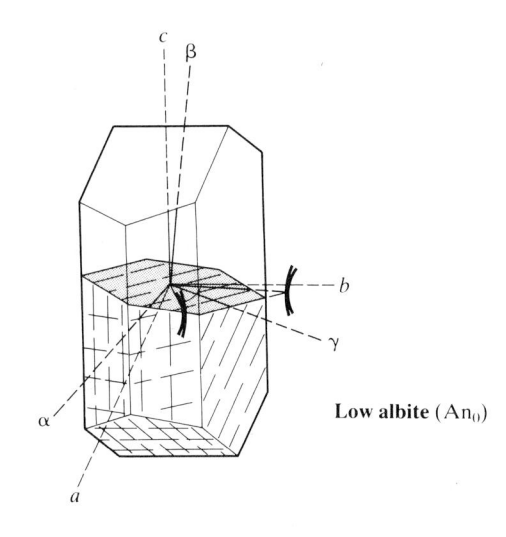

Low albite (An_0)

SILICATE MINERALS

Low microcline

showing cross-hatched twinning

600 μm

There is a continuous variation in RI from K- to Na-alkali feldspar, and general values for the end members are as follows:

	Ab	Or
$n\alpha$	1.527	1.518
$n\beta$	1.531	1.522
$n\gamma$	1.539	1.524
δ	0.012	0.006

$2V$ is variable in size or sign depending on composition and type; Figure 2.11 gives the full range of values. $2V$ values are 15°–40° sanidine, 42°–52° anorthoclase and 52°–54° high albite; all are biaxial negative, as is orthoclase (35°–50°) and microcline (66°–90°). Low albite is biaxial positive with $2V_y = 84°–78°$

OAP also varies; this variation is given in Figure 2.11 and in the various feldspar diagrams

$D = 2.56–2.63 \qquad H = 6–6\frac{1}{2}$

COLOUR Colourless with opaque patches if alteration to clay minerals has occurred.

HABIT Euhedral prismatic in high temperature porphyritic rocks to anhedral in plutonic intrusive rocks, although some porphyritic granites contain euhedral (high T) alkali feldspar phenocrysts; for example the granite from Shap Fell, Cumbria.

CLEAVAGE Two cleavages $\{001\}$ and $\{010\}$ meeting at nearly right angles on the (100) plane. Several partings may be present.

RELIEF Low, less than 1.54.

*ALTERATION Common, usually to clay minerals with K-feldspar altering as follows in the presence of a limited amount of water:

$$3 \text{ Or} + 2 \text{ H}_2\text{O} \rightarrow \text{illite} + 6 \text{ silica} + 2 \text{ potash}$$

or if excess water is present, then

$$2 \text{ Or} + 3 \text{ H}_2\text{O} \rightarrow \text{kaolin} + 4 \text{ silica} + 2 \text{ potash}$$

FELDSPAR GROUP

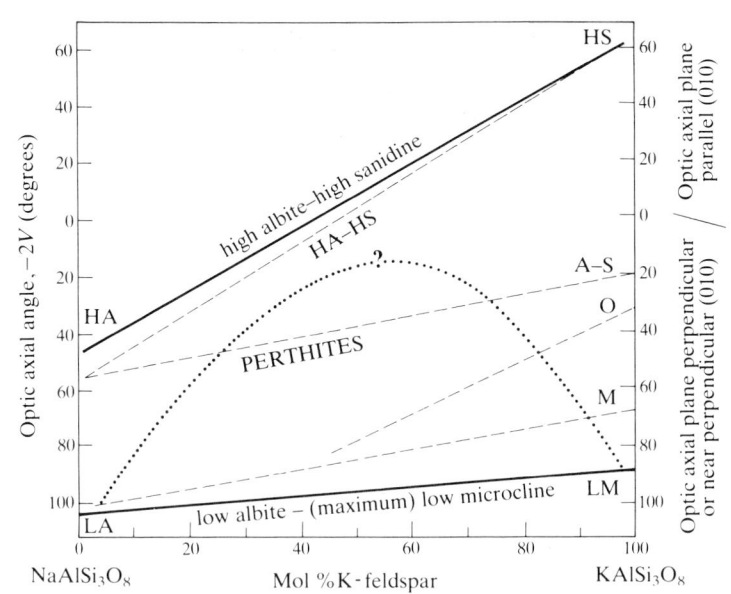

Figure 2.11 2V variation in alkali feldspars.

The clay minerals found occur as discrete tiny particles held within the feldspar crystal. As the amount of alteration increases, the clays increase both in amount and size to a point at which they are usually termed sericite. The sodium-rich alkali feldspars can alter in the same way that plagioclase feldspar does, with the clay mineral montmorillonite being formed

$$\text{Na-feldspar} + H_2O \rightarrow \text{montmorillonite} + Qz + \text{soda}$$

BIREFRINGENCE Low, first order grey or sometimes white are maximum colours.

INTERFERENCE FIGURE 2V is usually 40°–65° and negative, except for low albite which has a very large positive 2V. A single optic axis figure will be best, and for this an isotropic section is needed.

EXTINCTION Extinction angle measured to {010} cleavage trace varies depending on composition, and this is given in Figure 2.12.

*TWINNING Simple twins are common in the K-rich alkali feldspars, and the common monoclinic twin forms are shown in Figure 2.13 together with the two common triclinic twins. Low microcline, which is the common alkali feldspar type in sedimentary sandstones, metamorphic rocks and large plutonic acid intrusions, possesses a distinctive cross-hatched type of twinning in which both pericline and albite twin laws operate.

*PERTHITES These intergrowths of a Na-plagioclase in a K-feldspar host are always found in low temperature alkali feldspars.

*DISTINGUISHING FEATURES (a) *Sanidine–high albite*. 2V angles and extinction angles are small. Anorthoclase shows two sets of polysynthetic twins yielding a grid, or

SILICATE MINERALS

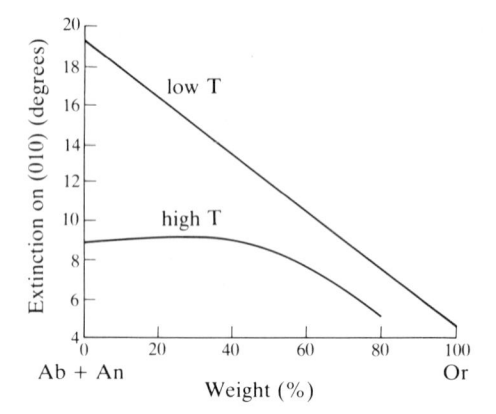

Figure 2.12 Feldspar extinction angles.

cross hatching, similar to microcline. Anorthoclase is confined to extrusive rocks, whereas microcline is found only in plutonic rocks. Microcline has a large $2V$ ($\sim 67°$) and is negative, but an interference figure is virtually impossible to obtain.

(b) *Orthoclase* is difficult to identify and can easily be mistaken for quartz, but its RIs are less than 1.54. Also it is frequently cloudy from alteration to clays, compared with quartz which is *always* unaltered and clear. Orthoclase is biaxial with a negative $2V_a$, which may increase in size if submicroscopic cryptoperthites exist, thus giving $2V_a$ values between orthoclase (max. 50°) and low albite ($\sim 90°$). Orthoclase is biaxial negative whereas nepheline is uniaxial negative, and orthoclase has a slightly larger $2V$ than its high temperature form sanidine.

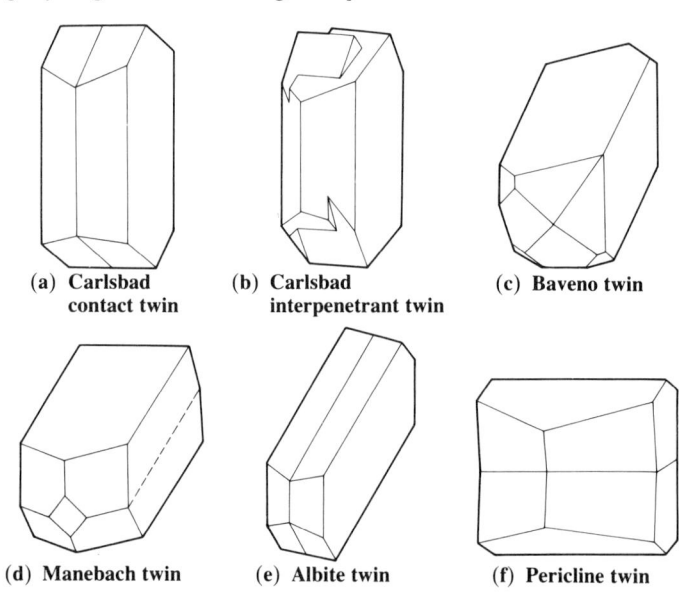

(a) **Carlsbad contact twin**

(b) **Carlsbad interpenetrant twin**

(c) **Baveno twin**

(d) **Manebach twin**

(e) **Albite twin**

(f) **Pericline twin**

Figure 2.13 Feldspar twin forms.

OCCURRENCE The alkali feldspars are essential constituents of alkali and acid igneous rocks, particularly syenites, granites and granodiorites, felsites and orthoclase porphyries and trachytes, rhyolites and dacites. Alkali feldspars are common in pegmatites, in hydrothermal veins and in high grade gneisses. Plutonic rocks contain orthoclase, microcline and perthites whereas extrusive rocks contain sanidine and other 'high' types. Perthite types can be correlated, based on the size of the exsolved phase, with decreasing temperature of formation and length of time of cooling period from cryptoperthites (< 5 nm obtained by X-ray diffraction data) found in hypabyssal rocks, to macroperthites (> 0.1 mm), which occur in igneous rocks formed from large plutonic intrusions.

Some pegmatites contain intergrowths of alkali feldspar and quartz called graphic intergrowth (because the quartz crystals resemble writing), and may be due either to simultaneous crystallisation of alkali feldspar and quartz or to the replacement of some of the alkali feldspar by quartz. Associations of feldspar, usually plagioclase but occasionally alkali feldspar, and quartz in which the feldspar and quartz are intercalated in 'stringer'-like textures, are called myrmekite. In Rapakivi granites, large crystals of alkali feldspar are mantled by plagioclase, and similar relationships occur in other orbicular granites. Potassium feldspars are common in high grade metamorphic rocks. It is a stable mineral at the highest grades when the breakdown of muscovite leads to the presence of K-feldspar, with sillimanite also being formed:

$$\text{muscovite} + \text{Qz} \rightarrow \text{K-feldspar} + \text{sillimanite}$$

These high grade metamorphic rocks containing K-feldspar include charnockites, sillimanite gneisses, migmatites, and granulites.

Orthoclase and microcline occur as detrital grains in terrigeneous arenaceous rocks; authigenic alkali feldspars, forming at low temperatures within sediments, have been recognised which are usually homogeneous (non-perthitic) and untwinned.

SILICATE MINERALS

Plagioclase feldspars

End members **high/low albite (Ab)** $NaAlSi_3O_8$ all triclinic
  high/low anorthite (An) $CaAl_2Si_2O_8$ $0.820\pm : 1.295\pm : 0.720\pm$
  $\alpha = 93°30'\pm, \beta = 116°\pm, \gamma = 88°–91°$

Low albite (An_0)

Labradorite (An_{50})

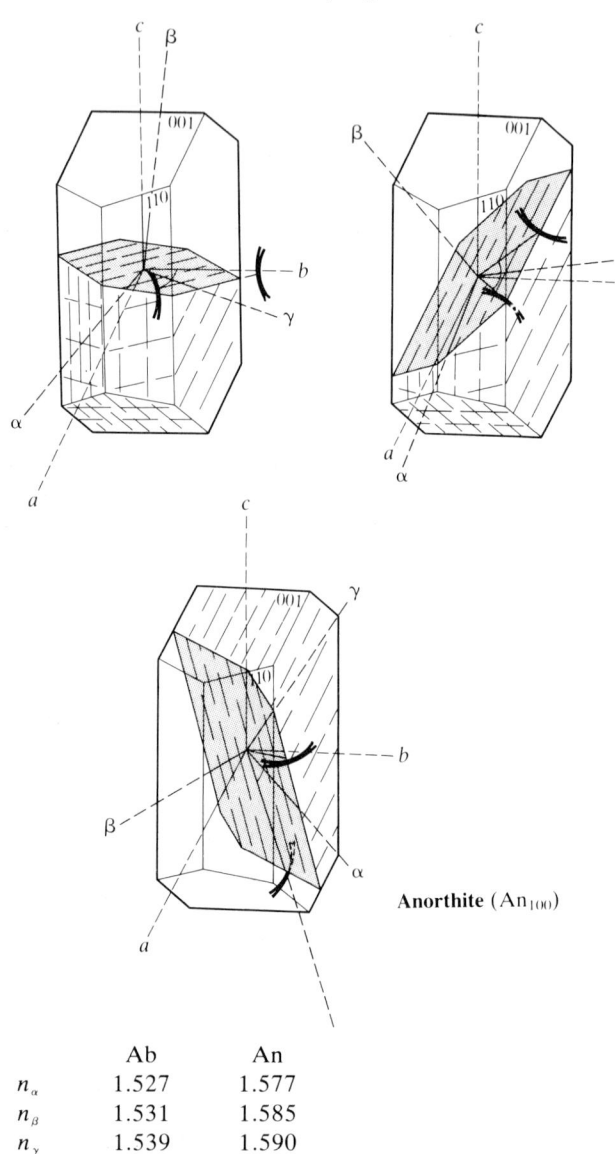

Anorthite (An_{100})

	Ab	An
n_α	1.527	1.577
n_β	1.531	1.585
n_γ	1.539	1.590
δ	0.012	0.013

All other plagioclases occur between these, with nomenclature:

An mol%	0	10	30	50	70	90	100
	Ab	oligoclase	andesine	labradorite	bytownite		An

There is continuous variation in the principal RIs of plagioclase feldspars from Ab to An, similar to the alkali feldspars, and this is given in Figure 2.14.

$2V$ is variable depending on the structural state and the composition, and Figure 2.15 shows the curves for variation of $2V$ for high (solid line) and low (dashed line) plagioclase feldspar.

$D = 2.62–2.76 \quad H = 6–6½$

COLOUR Colourless, occasionally near opaque because of clay development.
HABIT Subhedral prismatic in plutonic and hypabyssal rocks, to euhedral prismatic or tabular in extrusive rocks.
CLEAVAGE Similar to alkali feldspars. Two perfect cleavages $\{001\}$ and $\{010\}$ meeting at nearly right angles on the (100) plane. Several partings may occur.
RELIEF Low, but apart from albite all plagioclase feldspars have RIs greater than 1.54.
*ALTERATION Already partly dealt with under alkali feldspars, but it is worth repeating here that plagioclase feldspar alters either to montmorillonite with limited water available, or kaolin if excess water is available. This alteration may be the result either of late stage hydrothermal activity

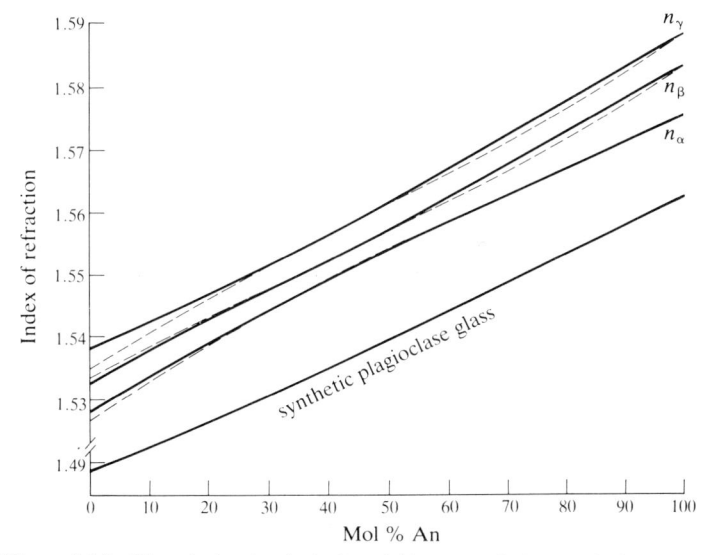

Figure 2.14 RI variation in plagioclase feldspars and glass made from these feldspars.

SILICATE MINERALS

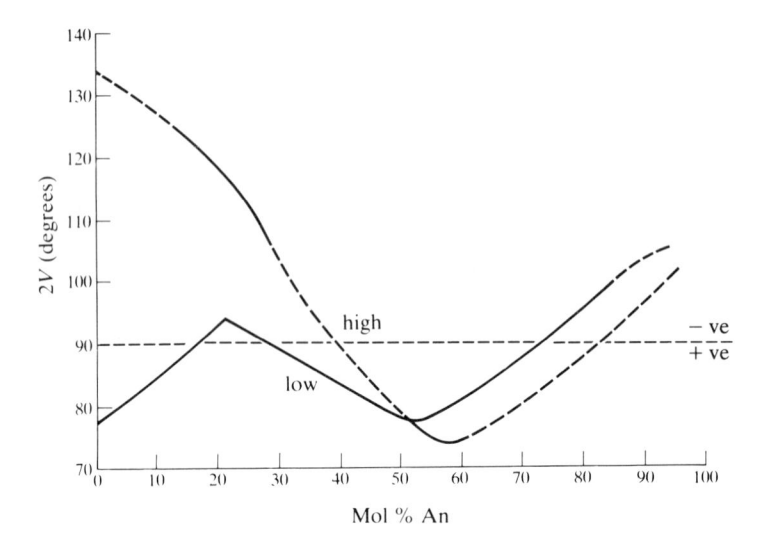

Figure 2.15
2V variation in plagioclase feldspars.

during solidification of the rock mass or of chemical weathering. Other minerals which may form from feldspars during late stage hydrothermal activity include the epidote mineral zoisite, or clinozoisite, which is produced during an alteration process called saussuritisation (see also the epidote group minerals).

BIREFRINGENCE Low, with interference colours varying from first order greys (Ab) to first order yellows (An).

INTERFERENCE FIGURE 2V is generally large and variable in sign with the data for high and low series shown in Figure 2.15. The OAP orientation varies from albite to anorthite, and Phillips and Griffen (1981) should be consulted for details.

*TWINNING Multiple twinning by the albite law is common in all plagioclase feldspars and is a *characteristic* feature. Albite twin lamellae, which tend to be parallel to the prism zone, often tend to be narrow in the Na-plagioclases and alternating narrow and broad in the Ca-plagioclases. Other twin laws which operate in plagioclases include Carlsbad (simple) and pericline (repeated), with combinations of twins common, such as Carlsbad–albite. Figure 2.16 gives a combined Carlsbad–albite twin, showing symmetrical extinction in each half of the twin. Sedimentary authigenic plagioclase feldspars (albites) and sodic plagioclases in low grade metamorphic rocks may be untwinned.

*EXTINCTION Composition of plagioclases may be determined by measuring the symmetrical extinction angles of albite twins measured on sections at right angles to the *a* crystallographic axis. Figure 2.17 gives full details of the variation in maximum extinction angle with composition, for both high (extrusive rocks) and low (hypabyssal and plutonic rocks) plagioclase feldspars. Combined Carlsbad–albite twins may also be used for determination of composition, but these combined twins are usually only

FELDSPAR GROUP

(a) Albite twin

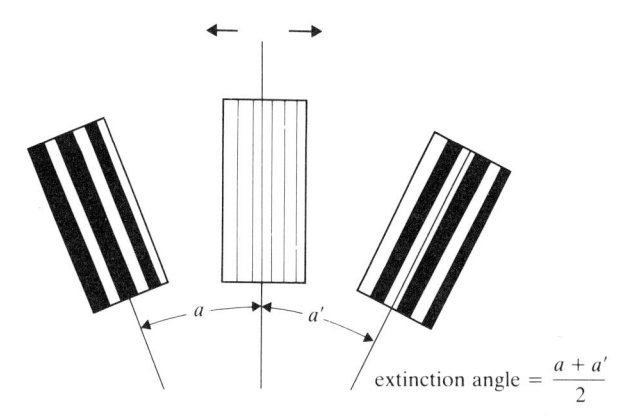

(b) Combined Carlsbad–albite twin

right-hand half

left-hand half

extinction angle = $\dfrac{a + a'}{2}$

smaller extinction angle = $\dfrac{a + a'}{2}$

larger extinction angle = $\dfrac{b + b'}{2}$

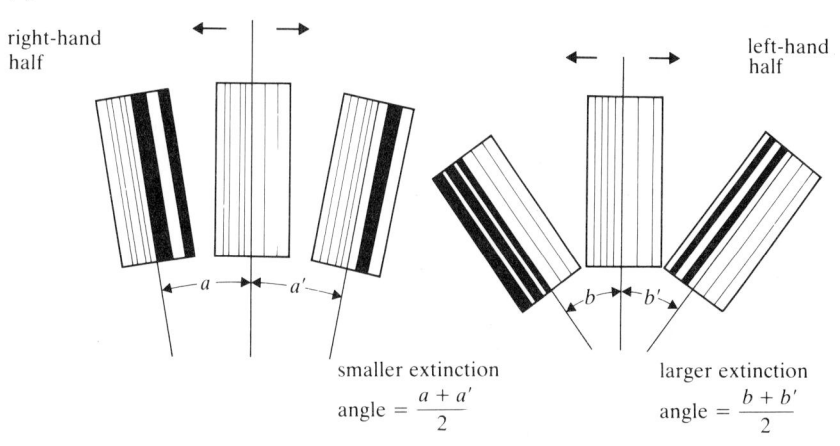

Figure 2.16 Measurement of extinction angles in (a) albite twin, (b) combined Carlsbad-albite twin.

found in ultrabasic igneous plutonic rocks (troctolites, peridotites) and some very basic extrusive types. Figure 2.18 shows the curves needed to obtain the composition of a combined twin in plutonic rocks, with each half of the Carlsbad twin being examined separately. Thus the smaller symmetrical extinction angle of one half of the twin is plotted along the ordinate and the larger symmetrical extinction angle of the other half of the twin is plotted on to the curves. The composition is then read off along the abscissa. No other twin types are commonly used in determining composition.

*ZONING Common in plagioclase feldspars from extrusive rocks. Normally the zoning shows up as a continuous change in composition from a calcium-rich core to a sodium-rich margin. In this case the composition should be

SILICATE MINERALS

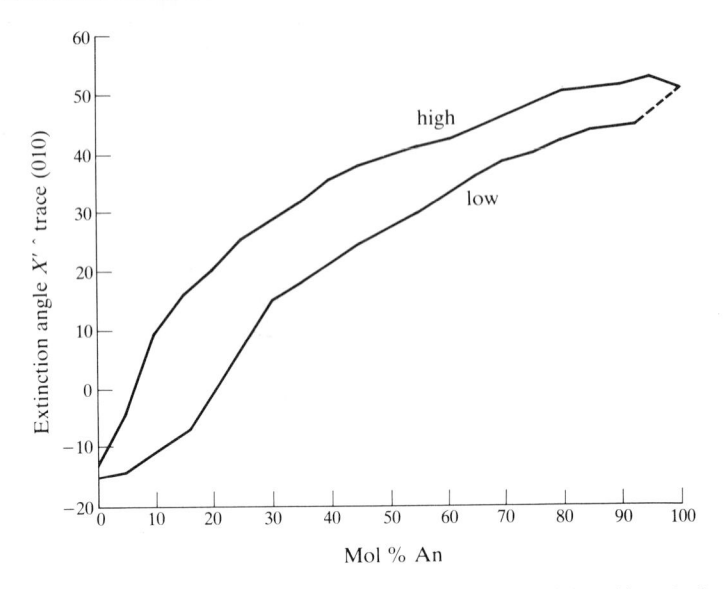

Figure 2.17 Maximum extinction angles for albite twins in high and low plagioclase feldspars.

given as, for example, An_{70} (core) to An_{52} (margin). If the zoning is reversed (sodium-rich core to calcium-rich margin) the precise composition variation should again be given, or if the zoning is oscillatory, where separated zones of equal extinction occur, an indication of the zonal variation should be given.

OCCURRENCE Plagioclase feldspars are almost always present in igneous rocks (with the exception of some ultramafic and ultra-alkaline types) and often comprise more than half the rock's total volume. Plagioclase varies in composition with the type of rock it is found in; thus bytownite occurs in ultrabasic rocks and labradorite in basic rocks, andesine is typical of intermediate rocks, and oligoclase is common in acid rocks.

In basic lavas, calcium-rich plagioclase feldspars occur both as phenocrysts and as constituents of the groundmass. In basic plutonic intrusions, layering and differentiation can occur, with feldspar-rich layers common. In these intrusions plagioclase may show a compositional range from An_{85} to An_{30} and is frequently zoned. The 'low' plagioclase found in plutonic rocks is often antiperthite, particularly in acid types, with exsolved alkali feldspar (K-feldspar). In other plutonic rocks, especially those with a long cooling history, peristerites (Fig. 2.10) may occur in which Schiller effects can be seen. The most basic intrusions may show either Bøggild intergrowths, caused by unmixing of two plagioclase components, or Huttenlocher intergrowths, which occur in bytownites in which two basic plagioclase components unmix; however, both these intergrowth types are rare and are rarely

FELDSPAR GROUP

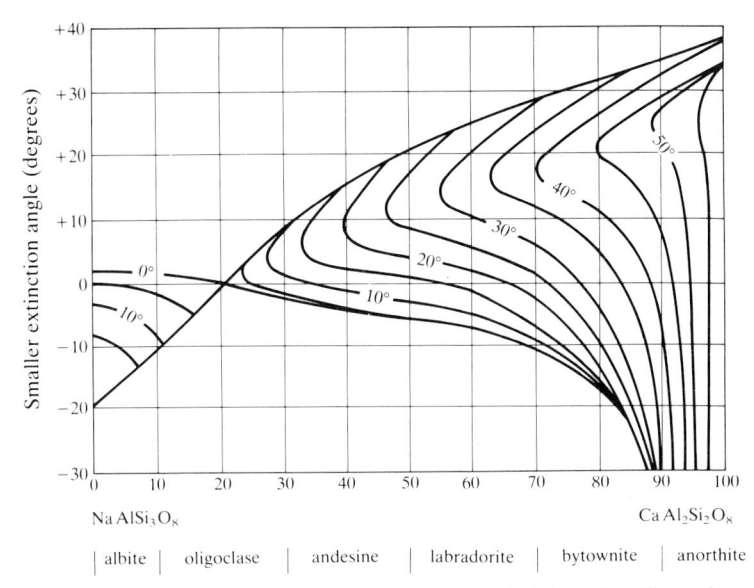

Figure 2.18 Graph for extinction angles of Carlsbad-albite twins. The extinction angle for an albite twin is measured in each half of a Carlsbad twin. The smaller angle is plotted along the ordinate and the larger angle *into* the nest of curves (see Fig. 2.16 for details of measurement of albite twin extinction angles). Thus, for example, a Carlsbad-albite twin with angles of extinction of 10° (the smaller) and 30° (the larger) has a composition of An_{60}. The negative ordinate values (below the horizontal line representing 0°) are needed for feldspars which have refractive indices of less than 1.54.

seen optically. Anorthosites contain plagioclase feldspars as the chief constituent, comprising well over 80% of the volume of the rock. The plagioclase composition varies from bytownite to andesine, although with any particular anorthosite intrusion the compositional range is quite small.

Pure albite is the typical feldspar of spilites, often with relict cores of a more anorthitic plagioclase. This may indicate a late stage magmatic or metasomatic process by which the original feldspar in the basalt becomes increasingly sodium rich, a process called albitisation. However, in some spilites the albite may be a primary crystallising mineral.

In metamorphic rocks, the composition of the plagioclase reflects the metamorphic grade of the rock, the plagioclase becoming more calcium-rich as the grade increases. Albite is the typical plagioclase of low grade regional rocks, with oligoclase occurring at garnet grade. In granulites and charnockites, andesine or rarely labradorite is the common plagioclase. Plagioclase feldspars do not occur in eclogites, the various feldspar components (Ca, Al etc.) entering either the clinopyroxenes or garnet phase present. Pure anorthite may occur in thermally metamorphosed calcareous rocks.

SILICATE MINERALS

In sedimentary rocks, apart from plagioclase occurring as detrital grains in many terrigeneous arenaceous rocks, albite may occur as an authigenic mineral in some sandstones, forming during sedimentation. Authigenic albite may be simply twinned but never shows lamellar twinning.

In addition to the normal feldspar group of minerals already discussed, some feldspars containing more than 2% BaO are termed barium feldspars. The most common barium feldspar ($BaAl_2Si_2O_8$), celsian, is similar in almost every property to the feldspars, especially orthoclase, except that it has greater RIs and a much higher density. This is rare and tends to be found associated with manganese deposits and stratiform barite deposits.

Feldspathoid family — Tektosilicates

The group of minerals termed 'feldspathoids' include those minerals which have certain similarities with the feldspars, particularly in their chemistry and structure. The main feldspathoid minerals given in detail here are:

leucite	$KAlSi_2O_6$
nepheline	$NaAlSiO_4$
sodalite	$Na_8Al_6Si_6O_{24}Cl_2$ or $6(NaAlSiO_4).2NaCl$

Other feldspathoid minerals include:

hauyne	$CaNa_6Al_6Si_6O_{24}.SO_4$ or $6(NaAlSiO_4).CaSO_4$ with S replacing SO_4
nosean	$Na_8Al_6Si_6O_{24}SO_4$ or $6(NaAlSiO_4).Na_2SO_4$
cancrinite	complex hydrated alkali aluminosilicate with CO_3, SO_4 and Cl groups included
kalsilite	$KAlSiO_4$

The feldspathoids are all silica deficient compared with the feldspars, and their occurrence is restricted to undersaturated alkali igneous rocks. Although analcime is a zeolite it is very closely associated with the feldspathoids and has been included here after sodalite.

Leucite $KAlSi_2O_6$ tetragonal
(pseudo-cubic) c/a 1.054

n = 1.508–1.511
δ = 0.001
If uniaxial, leucite is +ve but sign is virtually impossible to determine because of twinning.
D = 2.47–2.50 H = 5½–6

COLOUR	Colourless.
*HABIT	Usually euhedral crystals showing eight-sided sections.
CLEAVAGE	Very poor on $\{110\}$.
RELIEF	Low.
BIREFRINGENCE	Isotropic to very low.
*TWINNING	Repeated twinning on $\{110\}$ is always present and visible under crossed polars, as a type of cross hatching.
OTHER FEATURES	At high temperatures, leucite can contain sodium in the structure, which must be expelled at lower temperatures. Pseudo-leucite, a mixture of nepheline and feldspar, forms an intergrowth which completely replaces leucite in some rocks.
*OCCURRENCE	In potassium-rich basic extrusive rocks such as leucite–basanite, leucite–tephrite and leucitophyre, which are usually silica deficient. Pseudo-leucite may occur in some alkali basic plutonic rocks but mainly occurs in extrusive igneous rocks. At subsolidus temperatures, such as can be attained in plutonic intrusions, leucite breaks down to give nepheline and feldspar, which explains the absence of leucite and the presence of nepheline and feldspar assemblages in plutonic alkali rocks.

Nepheline $NaAlSiO_4$ hexagonal
(K may replace Na up to 25 per cent) c/a 0.838

n_o = 1.529–1.546
n_e = 1.526–1.542
δ = 0.003–0.005
Uniaxial −ve (crystal is length fast)
D = 2.56–2.66 H = 5½–6

COLOUR	Colourless.
HABIT	Usually anhedral, occurring in the interstices between minerals. Occasionally found as small exsolved 'blebs' within feldspars, particularly K feldspars. Euhedral crystals have a hexagonal outline.
CLEAVAGE	$\{1010\}$ imperfect prismatic cleavage, and poor basal $\{0001\}$ cleavage.
RELIEF	Low.
ALTERATION	Nepheline may alter to zeolites such as natrolite $Na_2Al_2Si_3O_{10}.2H_2O$ or analcime, and to feldspathoids such as sodalite e.g.

$$2NaAlSiO_4 + SiO_2 + 2H_2O \rightarrow Na_2Al_2Si_3O_{10}.2H_2O \text{ (natrolite)}$$
$$NaAlSiO_4 + SiO_2 + H_2O \rightarrow NaAlSi_2O_6.H_2O \text{ (analcime)}$$

by the addition of silica, water and other volatiles (chlorine in the case of sodalite). Nepheline commonly alters to cancrinite (a complex hydrated silicate with sulphate, carbonate and chloride).

*BIREFRINGENCE	Low, first order greys. Small inclusions occur within nepheline and give the crystal a 'night sky' effect under crossed polars.
TWINNING	Rare.

SILICATE MINERALS

*OCCURRENCE A characteristic primary crystallising mineral of alkali igneous rocks. Nepheline is an essential constituent of silica-deficient nepheline–syenites and may occur in volcanic rocks where it is associated with high-temperature feldspars.

Nepheline may be metasomatic in origin, formed by the reaction of alkali-rich magmatic fluids with country rocks. Nepheline may occur in basic rocks near their contact with carbonate rocks; and alkali dolerites with interstitial nepheline have also been described.

Another mineral associated with nepheline is kalsilite ($KAlSiO_4$) which forms a limited solid solution with $NaAlSiO_4$. Up to 25% of the nepheline molecule can be replaced with kalsilite, although the amount of solid solution increases with increasing temperature, up to a maximum of 70% at 1070 °C. Kalsilite is not found in plutonic igneous rocks but is found in the ground mass of some K-rich lavas.

Sodalite $Na_8Al_6Si_6O_{24}Cl_2$ cubic

$n = 1.483–1.487$
$D = 2.27–2.88 \quad H = 5\frac{1}{2}–6$

COLOUR Colourless, sometimes pale blue or pink.
HABIT Usually anhedral but occasional eight-sided anhedral crystals occur.
CLEAVAGE Poor $\{110\}$.
RELIEF Moderate (much less than CB).
* Sodalite and analcime are virtually indistinguishable in thin section. The type of rock is as good a guide as any as to which mineral it is.
OCCURRENCE Found in nepheline–syenites in association with nepheline and fluorite. It occurs in metasomatised calcareous rocks near alkaline intrusions.

Nosean
Hauyne
Cancrinite
Two related minerals, nosean ($Na_8Al_6Si_6O_{24}SO_4$) and hauyne, which has a similar composition to nosean with some S and Ca in the structure, have virtually identical optical properties. Nosean is distinguished by having a dark border around each crystal and usually occurs in silica-poor alkaline rocks such as phonolites and leucitophyres. Hauyne is indistinguishable from sodalite and also occurs in phonolites and other undersaturated rocks, but is usually accompanied by sulphide minerals such as pyrite. Cancrinite shows higher birefringence than nepheline (first order yellows and reds), is uniaxial negative, and is a characteristic mineral in similar rocks to those in which nepheline occurs. Cancrinite may be an alteration product of nepheline.

Analcime (Analcite) $NaAlSi_2O_6.H_2O$ cubic

$n = 1.479–1.493$
$D = 2.34–3.29 \quad H = 5\frac{1}{2}$

COLOUR Colourless.
HABIT Anhedral crystals filling interstices between mineral grains.

CLEAVAGE Poor {001} fracture.
RELIEF Moderate (less than 1.54).
*OCCURRENCE It has been suggested that the analcime in some intermediate and basic rocks such as dolerite, teschenites and essexites (porphyritic alkali gabbros) is primary, but recent work has questioned this. In volcanic rocks analcime occurs in some basalts as a primary mineral, but most commonly as a late stage hydrothermal mineral crystallising in vesicles and found with zeolites – especially thomsonite and stilbite.

Analcime occurs as an authigenic mineral in sandstones, again associated with the zeolites laumontite and heulandite. Analcime has also been found in pyroclastic rocks.

Garnet group   Nesosilicates

$X_3Y_2Si_3O_{12}$ cubic

			n	D	
Almandine	(X = Fe)	(Y = Al)	1.830	4.318	
Pyrope	(X = Mg)	(Y = Al)	1.714	3.582	
Grossular	(X = Ca)	(Y = Al)	1.734	3.594	$H = 6-7\frac{1}{2}$
Spessartine	(X = Mn)	(Y = Al)	1.800	4.190	
Andradite	(X = Ca)	(Y = Fe)	1.887	3.859	

Al^{3+} occupies the Y position in the above minerals, but Cr^{3+}, Fe^{3+}, and Ti^{4+} as well as REEs may also substitute in garnet minerals in this position. Solid solutions covering a wide range are possible within the garnet group of minerals, with a corresponding wide range of optical properties – RI varies from 1.890 to 1.710, and density from 4.32 to 3.59. A variety called hydrogrossular, which contains hydroxyl groups, may show an RI as low as 1.68 and a density down to 3.1.

COLOUR Colourless, pale brown or pale pink, dark green or brown.
*HABIT Euhedral crystals of garnet showing six sided or eight sided sections are common.
*CLEAVAGE No cleavage exists but crystals often fractured.
*RELIEF High to very high; sections always appear to have a rough surface.
ALTERATION Fe-Mg bearing garnets typically alter to chlorite (by hydrothermal reaction), with quartz being produced in the reaction.

Garnet is isotropic but sometimes hydrogrossular in particular may be anisotropic and zoned.

*OTHER FEATURES Garnet occasionally exhibits compositional zoning, and in metamorphic medium grade rocks garnet frequently contains inclusions of quartz and micas. Some metamorphic garnets contain inclusions defining an earlier fabric (e.g. schistosity) which has been trapped within the garnet.

OCCURRENCE Garnet occurs mainly in metamorphic rocks. Almandine (containing some Mg) is the main garnet of middle grade metamorphism of pelitic rocks and igneous rocks. Almandine forms by progressive metamorphic reactions involving chlorite, and by a reaction between quartz and

SILICATE MINERALS

staurolite which gives garnet and kyanite. In high heat flow, low P metamorphism garnet may form from cordierite breakdown if the chemistry is correct. In thermal aureoles in, and regional metamorphism of, impure limestones, grossular forms; and spessartine forms during the metamorphism of Mn-rich rocks. Andradite occurs in thermally metamorphosed impure calcareous sediments, and particularly in metasomatic skarns.

Pyrope garnet is an essential constituent of some ultrabasic igneous rocks, especially garnet–peridotite and other similar types derived from the Earth's upper mantle. High grade metamorphic rocks (very high P and T) called eclogites have garnet of almandine–pyrope composition as an essential constituent, along with the pyroxene omphacite.

Garnet also occurs as a detrital mineral in sands.

Humite group

$n\text{Mg}_2\text{SiO}_4\text{Mg}(\text{OH},\text{F})_2$

There are four members of the group; norbergite (with $n = 1$), chondrodite ($n = 2$), humite ($n = 3$) and clinohumite ($n = 4$).

Nesosilicates

orthorhombic
$0.463 : 1 : \begin{matrix} 0.855 \text{ (No)} \\ 2.057 \text{ (Hu)} \end{matrix}$

monoclinic
$2.170 : 1 : 1.663$ (Ch)
$0.462 : 1 : 1.332$ (Cl)

Chondrodite
section parallel to 010

Humite
(OAP section)
section parallel to 001 (nordbergite is similar)

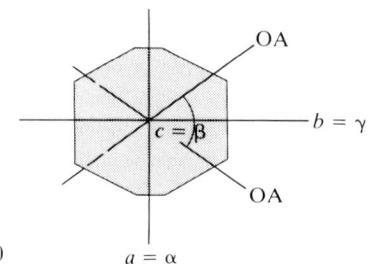

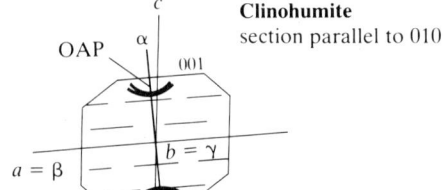

Clinohumite
section parallel to 010

HUMITE GROUP

	n_α	n_β	n_γ	δ
Norbergite	1.561–1.567	1.567–1.579	1.587–1.593	0.026
Chondrodite	1.592–1.643	1.602–1.655	1.619–1.675	0.025–0.037
Humite	1.607–1.643	1.619–1.653	1.639–1.675	0.028–0.036
Clinohumite	1.623–1.702	1.636–1.709	1.651–1.728	0.028–0.045

Variation in RI is caused by Fe^{2+} and Ti^{4+} entering structure.

	$2V_\gamma$	D	
Norbergite	44°–50° +ve	3.15–3.18	$H = 6\frac{1}{2}$
Chondrodite	64°–90° +ve	3.16–3.26	
Humite	65°–84° +ve	3.20–3.32	$H = 6$
Clinohumite	52°–90° +ve	3.21–3.35	

*COLOUR   Pale yellow or yellow.

PLEOCHROISM   Norbergite, chondrodite and humite have α pale yellow, β colourless or rarely pale yellow, and γ colourless. Clinohumite has α golden yellow, and β, γ pale yellow.

HABIT   Anhedral masses of crystals usually occur. Occasionally large subhedral porphyroblasts can be present.

CLEAVAGE   Basal $\{001\}$ usually present.

RELIEF   Moderate to high (clinohumite).

ALTERATION   All the humite minerals alter to serpentine or chlorite, as follows:

$$Mg_2SiO_4 \cdot Mg(OH,F)_2 + H_2O + SiO_2 \rightarrow Mg_3Si_2O_5(OH,F)_4$$

*BIREFRINGENCE   Moderate to high (clinohumite) giving maximum upper second order interference colours (lower third order, clinohumite).

INTERFERENCE FIGURE   Single optic axis figure yields positive $2V$ which varies in size in different humite minerals (see above).

EXTINCTION   Norbergite and humite have straight extinction on cleavage (both orthorhombic), whereas chondrodite has $\alpha\hat{}cl = 3°$ to $12°$ and clinohumite $\alpha\hat{}cl = 0°$ to $4°$ (both monoclinic forms).

TWINNING   Simple or multiple twinning on $\{001\}$ in monoclinic forms.

ZONING   Common, shown by colour intensities.

DISTINGUISHING FEATURES   Similar to olivine except for pale yellow colour with moderate interference colours. Yellow colour in olivine normally implies alteration to serpentine with low or anomalous interference colours. Olivine $2V$ is very large and usually negative. Staurolite has higher RIs, lower birefringence and occurs in schists. Individual humite group members are difficult to distinguish from each other.

OCCURRENCE   The humite minerals have a restricted occurrence, being found in contact metamorphosed and metasomatised limestones and dolomites, near acid or alkaline intrusions.

SILICATE MINERALS

Melilite group — Sorosilicates

$(Ca,Na)_2(Mg,Al)(Al,Si)_2O_7$
tetragonal
$c/a = 0.65$

This group comprises a series from the Ca,Al end member gehlenite to the Ca,Mg end member åkermanite. The mineral melilite occupies an intermediate position with Na and Fe^{2+} included as well as Ca, Mg and Al.

n_o = 1.670 (Geh)–1.632 (Åk)
n_e = 1.658 (Geh)–1.640 (Åk)
δ = 0.012–0.000 (Geh); 0.000–0.008 (Åk)
Uniaxial +ve (Geh), −ve (Åk)
D = 3.04–2.944 H = 5–6

COLOUR — Colourless, variable in browns.
PLEOCHROISM — Absent from all except thick sections of melilite with o golden brown, e colourless.
*HABIT — Melilite group minerals usually appear as subhedral to euhedral crystals, tabular on the basal plane.
CLEAVAGE — {001} moderate, {110} poor.
RELIEF — Moderate.
ALTERATION — Melilite alters to fibrous masses of cebollite $Ca_5Al_2Si_3O_{14}(OH)_2$ or prehnite $Ca_2Al(AlSi_3)O_{10}(OH)_2$; the latter reaction is as follows:

$$Ca_2Al_2SiO_7 + H_2O + 2SiO_2 \rightarrow Ca_2Al_2Si_3O_{10}(OH)_2$$

Other minerals like calcite, zeolite, garnet have been reported as alteration products from melilite breakdown.

*BIREFRINGENCE — Low to vey low, anomalous interference colours may be seen, with dark blue instead of first order grey.
INTERFERENCE FIGURE — Difficult to obtain, with poorly defined isogyres on a pale grey field.
ZONING — Oscillatory zoning common.
DISTINGUISHING FEATURES — Very low birefringence and anomalous interference colours are distinctive, as is the limited occurrence.
*OCCURRENCE — Melilite occurs in the groundmass of some silica-poor calcium-rich extrusive basalts (melilitites). Melilite may form under high T low P metamorphism (sanidinite facies) at the contact between silica-poor magmas and carbonate sediments.

Mica group — Phyllosilicates

The micas contain sheets of cations such as Fe, Mg or Al (the octahedral sheets), which are linked to two sheets of linked (SiO_4) tetrahedra; the tetrahedral sheets and the mica minerals belong to the 2:1 layer silicates because of this ratio of tetrahedral to octahedral sheets. The complete 'sandwich unit' is linked to another similar unit by weakly bonded

MICA GROUP

monovalent cations K^+ or Na^+, and the perfect mica cleavage occurs along these planes of K^+ or Na^+ ions. Some oxygen ions are replaced by hydroxyl ions. Thus the general formula is:

$X_2Y_{4-6}Z_8O_{20}(OH,F)_4$ with X = K, Na (Ba, Rb, Cs, Ca)
Y = Mg,Fe,Al (Mn, Li, Ce, Ti, V, Zn, Co, Cu, V)
Z = Si, Al (Ti, Ge)

The principal mica group minerals described in detail are phlogopite, biotite and muscovite. Other micas which may be encountered in rocks include lepidolite (Li-bearing, colourless) found in granite pegmatites, paragonite (Na-bearing, colourless) found in some sodium-rich metamorphic rocks, and glauconite (K, Na and Ca-bearing and iron rich) – a complex mica, green in colour and pleochroic in greens. It is formed as an authigenic mineral in sediments – limestones, sandstones and siltstones – and is produced in reducing conditions. All these minerals have properties similar to the micas described (particularly biotite).

Phlogopite $K_2(Mg,Fe)_6Si_6Al_2O_{20}(OH,F)_4$

monoclinic
pseudo-hexagonal
$0.576:1:1.109; \beta = 100°00'$

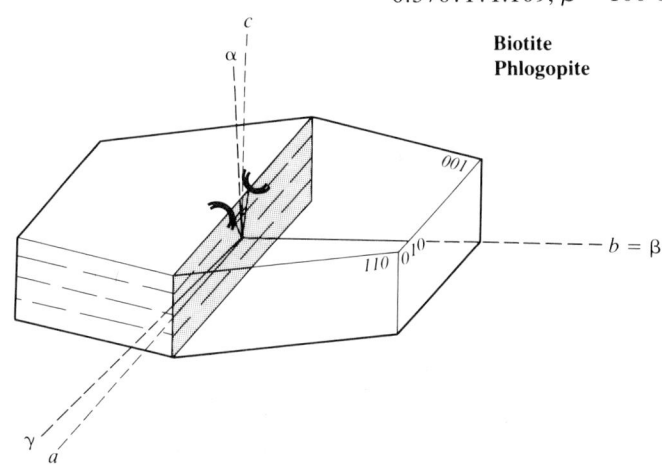

Biotite
Phlogopite

$n_\alpha = 1.530–1.590$
$n_\beta = 1.557–1.637$
$n_\gamma = 1.558–1.637$
$\delta = 0.028–0.049$

RIs increase with increasing iron content, although Mn and Ti presence will also increase RIs. RIs will decrease with increasing F content

$2V_\alpha = 0–15°$ −ve
OAP is parallel to (010)
D = 2.76–1.90   H = 2½

SILICATE MINERALS

COLOUR	Pale brown, colourless.
*PLEOCHROISM	Weak, with pale colours; α yellow, β and γ brownish red, green, deeper yellow.
*HABIT	Small tabular crystals common, frequently subhedral.
*CLEAVAGE	Perfect $\{001\}$ cleavage.
RELIEF	Low to moderate.
*BIREFRINGENCE	High, third order colours commonly present; weak body colours only slightly mask the interference colours.
*INTERFERENCE FIGURE	A basal section shows a good Bx_a figure with a very small $2V$.
EXTINCTION	Usually straight, but slight angle $\beta\hat{}$cleavage = 5° (max) on (010) face may occur.
TWINNING	Rare on $\{310\}$, seen on cleavage section.
OTHER FEATURES	Reaction rims may occur in phlogopites found in kimberlite intrusions.
*OCCURRENCE	Found in metamorphosed impure magnesian limestones where phlogopite forms by reactions between the dolomite and either potash feldspar or muscovite. Phlogopite is a common constituent of kimberlite, occurs in many leucite-bearing rocks, and is a minor constituent of ultramafic rocks.

Biotite $K_2(Mg,Fe)_{6-4}(Fe^{3+},Al,Ti)_{0-2}Si_{6-5}Al_{2-3}O_{20}(OH,F)_4$ monoclinic
(identical to phlogopite)

$n_\alpha = 1.565-1.625$
$n_\beta = 1.605-1.696$
$n_\gamma = 1.605-1.696$
$\delta = 0.040-0.080$
$2V_\alpha = 0-25°$ $-ve$
OAP is parallel to (010)
$D = 2.7-3.3$ $H = 2\frac{1}{2}$

COLOUR	Brown or yellowish; occasionally green.
*PLEOCHROISM	Common and strong with α yellow, β and γ dark brown. Note that pleochroism cannot be detected in a basal section, and that a prism section showing a cleavage is best.
*HABIT	Tabular, subhedral hexagonal plates.
*CLEAVAGE	$\{001\}$ perfect.
RELIEF	Moderate.
*ALTERATION	Common in rocks which have undergone hydrothermal alteration. The biotite alters to chlorite with potash being released in the reaction. The reverse reaction occurs during progressive metamorphism, with chlorite changing in composition as biotite forms at higher temperatures.
*BIREFRINGENCE	High to very high but masked by body colour. Note that the birefringence of basal sections of all micas is virtually zero since the n_β and n_γ values are almost equal, and n_β and n_γ lie in the (001) plane (i.e. the basal section).

MICA GROUP

*INTERFERENCE FIGURE A basal section gives a perfect Bx_a figure with a very small $2V$, although the strong body colour may tend to mask the colour that determines the sign of the mineral.

EXTINCTION Nearly straight on cleavage. A speckled effect is seen near the extinction position, which is very characteristic of micas.

TWINNING Extremely rare, similar to phlogopite twins.

OCCURRENCE Common mineral in a variety of rocks. Biotite occurs in most metamorphic rocks which have formed from argillaceous sediments. It forms after chlorite and is a constituent of regional metamorphic rocks existing up to very high grades, although its composition changes with the metamorphic grade. These changes may be accompanied by changes in colour, due to Mg and Ti increasing in amount in the mineral.

Biotites are primary crystallising minerals in acid and intermediate plutonic igneous rocks, and in some basic rocks. Biotite is not common in acid and intermediate extrusive and hypabyssal rocks.

Biotite is a common mineral in clastic arenaceous sedimentary rocks but is very prone to oxidation and degradation.

The type of mica depends on its Mg:Fe ratio. Thus:

phlogopite is a mica with Mg^{2+} between 100 and 70%
biotite is a mica with Mg^{2+} between 60 and 20% (and increasing R^{3+})
siderophyllite is a mica with $Mg^{2+} < 10\%$ (R^{3+} is Al^{3+})
lepidomelane is a mica with $Mg^{2+} < 10\%$ (R^{3+} is Fe^{3+})
annite is a mica with Mg^{2+} 0% (R^{3+} is zero).

SILICATE MINERALS

Muscovite $K_2Al_4Si_6Al_2O_{20}(OH,F)_4$ monoclinic
$0.574:1:2.217, \beta = 95°30'$

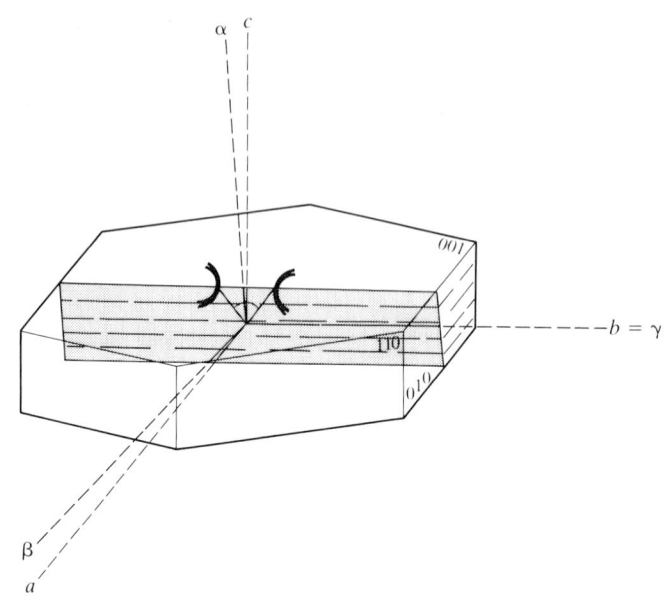

$n_\alpha = 1.552–1.574$
$n_\beta = 1.582–1.610$
$n_\gamma = 1.587–1.616$
$\delta = 0.036–0.049$
$2V_\alpha = 30°–47°$ −ve
OAP is perpendicular to (010)
$D = 2.77–2.88 \quad H = 2½ +$

COLOUR	Colourless.
HABIT	Thin platy crystals; occasionally in aggregates. When the aggregates consist of very fine grained crystals the mineral is called sericite.
*CLEAVAGE	{001} perfect.
RELIEF	Low to moderate (if iron enters the structure).
ALTERATION	Absent.
*BIREFRINGENCE	High, upper second order to third order. Basal sections show a low birefringence of first order grey which has a characteristic speckled appearance.
*INTERFERENCE FIGURE	A basal section gives an excellent Bx_a figure which just fits into the field of view; i.e. both isogyres will just be seen, indicating a 2V size of 40° ±.
EXTINCTION	Straight on cleavage.
TWINNING	Similar to phlogopite but not observable.
OCCURRENCE	Muscovite is found in low grade metamorphic pelitic rocks where it forms from pyrophyllite or illite. Muscovite remains in these rocks as the

grade increases, eventually becoming unstable at highest temperatures (> 600 °C) when the following reaction occurs:

muscovite + quartz → K-feldspar + sillimanite

Muscovite occurs in acid igneous plutonic rocks where it is a late crystallising component, and it occurs in some detrital (clastic) arenaceous sedimentary rocks.

Olivine group   Nesosilicates

Forsterite (Fo) Mg_2SiO_4 orthorhombic
Fayalite (Fa) Fe_2SiO_4
0.467 : 1 : 0.587
0.458 : 1 : 0.579

olivine group

n_α = 1.635 (forsterite)–1.824 (fayalite)
n_β = 1.651   −1.864
n_γ = 1.670   −1.875
δ = 0.035   −0.051
$2V_\gamma$ = 82°–134°, i.e. $2V_\gamma$ 82°–90° +ve, $2V_\alpha$ 90°–46° −ve
OAP is parallel to (001)
D = 3.22 (forsterite) – 4.39 (fayalite) $H = 6½$

SILICATE MINERALS

COLOUR   Usually colourless, but may appear pale yellow if Fe^{2+} content very high.

PLEOCHROISM   Extremely rare; only Fa olivines may show α and β pale yellow, with γ yellow.

HABIT   In igneous rocks olivine ranges from anhedral (plutonic rocks) to subhedral (extrusive rocks). A rough six-sided crystal shape can be seen in basalts where the olivine occurs as phenocrysts.

CLEAVAGE $\{010\}$ poor with rare $\{100\}$ imperfect fracture.

RELIEF   Variable depending upon composition; forsterite is moderate to high and fayalite very high.

*ALTERATION   Olivine is very susceptible to hydrothermal alteration, low grade metamorphism and the effects of weathering. Mixtures of products are produced, the most common being serpentine, chlorite, talc, carbonates, various iron oxides and iddingsite and bowlingite (which appear to be mixtures of hydrated iron oxides including goethite and hematite). The most common reaction is that of olivine altering to serpentine:

$$3 \text{ olivine} + \text{water} + \text{silica} \rightarrow \text{serpentine}$$

serpentine may break down further to talc and carbonates.

Some alteration of olivine is usually present along irregular basal fractures and poor cleavages, and complete alteration can occur with the olivine crystal totally replaced by serpentine, often with a release of iron ores.

*BIREFRINGENCE   High, with maximum interference colour lower third order (for iron-rich olivines).

*INTERFERENCE FIGURE $2V$ is generally very large, and a single isogyre should be examined from an isotropic section. Such a grain will not appear black on rotation but will appear greyish, owing to the high relief and dispersion of olivine. The optic axis figure will show an isogyre in which the direction of curvature is difficult to determine if $2V$ is very large ($> 80°$). Olivines from most basic igneous rocks have a compositional range of $Fo_{85}Fa_{15}$ to $Fo_{50}Fa_{50}$, which represents a $2V$ range of from 90° to 75° (Fig. 2.19).

EXTINCTION   Straight on poor $\{010\}$ cleavage or prism face.

TWINNING   Rare in most olivines. Sometimes broad deformation lamellae parallel to (100) occur in ultramafic igneous rocks.

OTHER FEATURES   Zoning is occasionally present but is not a diagnostic feature. Mg-rich olivine may contain minor amounts of trivalent ions which may be exsolved as tiny inclusions of chromite or magnetite.

OCCURRENCE   Mg-rich olivine is an essential mineral in most ultrabasic igneous rocks – dunites, peridotites and picrites. Olivine is also common in many basic igneous rocks such as gabbros, dolerites and basalts, and more Fe-rich olivine may occur in some undersaturated igneous rocks such as larvikites, teschenites and alkali basalts. Iron-rich olivine occurs in some ferrogabbros, trachytes and syenites. Olivine can occur in metamorphosed basic igneous rocks and metamorphosed dolomitic limestones, in which the olivine is nearly pure forsterite.

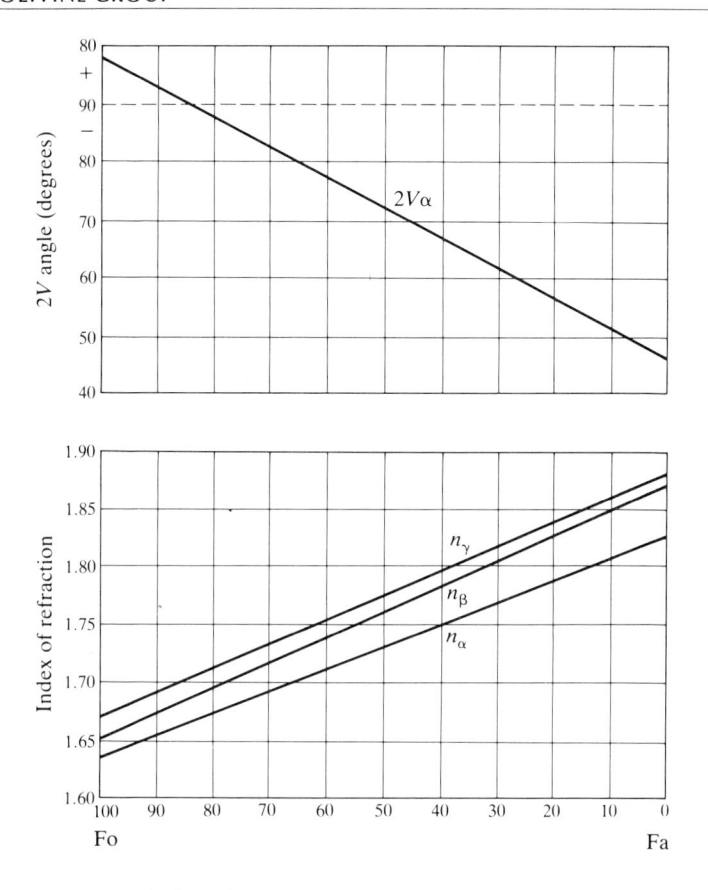

Figure 2.19 Variation of $2V$ angle and indices of refraction for the forsterite–fayalite (olivine) series (after Bowen & Schairer 1935).

Olivines may show reaction rims against plagioclase crystals (called kelyphitic margins or corona structures) in metamorphosed basic igneous rocks or in some ultrabasic troctolites, where an olivine or olivine plus serpentine kernel is surrounded by successive rims of pyroxene and garnet or spinel and amphibole. In basic igneous rocks containing Mg-rich olivine, quartz is never present since it would have combined with the olivine in the crystallising magma to give more orthopyroxene; however, in some later stage iron-rich igneous rocks (syenogabbros, ferrogabbros), iron-rich olivine coexists with quartz.

SILICATE MINERALS

Pumpellyite   Sorosilicate

Pumpellyite $Ca_2Al_2(Mg,Fe^{2+},Fe^{3+},Al)(SiO_4)(Si_2O_7)(OH)_2(H_2O.OH)$ monoclinic
$1:1:0.842, \beta = 97°\,36'$

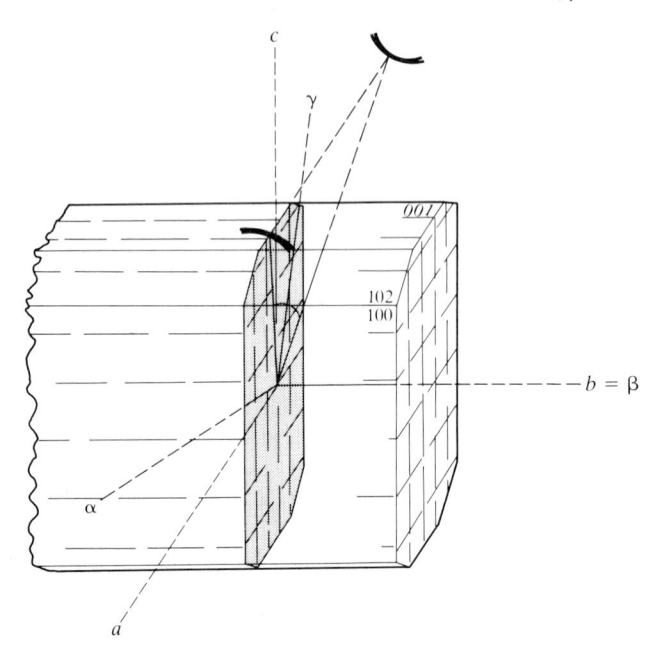

$n_\alpha = 1.674–1.748$
$n_\beta = 1.675–1.754$ } RIs vary with increasing iron content
$n_\gamma = 1.688–1.764$
$\delta = 0.002–0.022$
$2V_\gamma = 10°–85°$ +ve
OAP is parallel to (010)
$D = 3.18–3.23 \qquad H = 6$

*COLOUR	Green or yellow.
*PLEOCHROISM	α colourless, pale yellow, β pale green, yellow, γ colourless yellowish brown.
HABIT	Aggregates of acicular or fibrous crystals, often in rosettes, are common.
CLEAVAGE	$\{001\}$ and $\{100\}$ present.
RELIEF	High.
ALTERATION	Uncommon.
BIREFRINGENCE	Low to moderate with variable interference colours from first order greys to second order blues.
*INTERFERENCE FIGURE	Highly variable $2V_\gamma$ from 10° to 85° depending on composition, always positive. Near basal sections give a reasonable Bx_a figure, but a single optic axis figure is needed for iron-rich varieties.

EXTINCTION	Oblique on cleavage with $\gamma\hat{}cl$ and $\alpha\hat{}cl$ both $\approx$ 4° to 30° seen on an (010) section.
TWINNING	Sector twinning common on $\{001\}$ and $\{100\}$.
DISTINGUISHING FEATURES	Pumpellyite is similar to epidote but has a deeper colour and is biaxial positive (epidote is biaxial negative).
OCCURRENCE	Pumpellyite is common in low grade regionally metamorphosed schists and is characteristically developed in glaucophane–schist facies rocks.

Pumpellyite may form by hydrothermal action in low temperature veins, amygdales and altered mafic rocks.

Pyroxene group   Inosilicates

Orthorhombic pyroxenes (usually abbreviated to opx) and monoclinic pyroxenes or clinopyroxenes (abbreviated to cpx) occur. A wide range of compositions is available. The general formula may be expressed as:

$$X_{1-n}Y_{1+n}Z_2O_6$$

where $X = Ca, Na$, $Y = Mg, Fe^{2+}, Ni, Li, Fe^{3+}, Cr, Ti$, and $Z = Si, Al$.

In orthopyroxenes n is approximately 1 and virtually no trivalent or monovalent ions are present; thus the formula reduces to $Y_2Z_2O_6$ or $(Mg,Fe,Mn)_2(Si,Al)_2O_6$. In clinopyroxenes n varies from 0 to 1, and the elemental substitutions must be such that the sum of the charges in X, Y and Z balances the six O^{2-} ions.

Some pyroxenes have compositions that can be expressed in terms of $CaSiO_3$–$FeSiO_3$–$MgSiO_3$ end members, and Figure 2.20 shows the main *monoclinic* pyroxene minerals occurring in this field of composition. The principal change occurring in this group is a variation in symmetry

Figure 2.20 Composition diagram for pyroxenes.

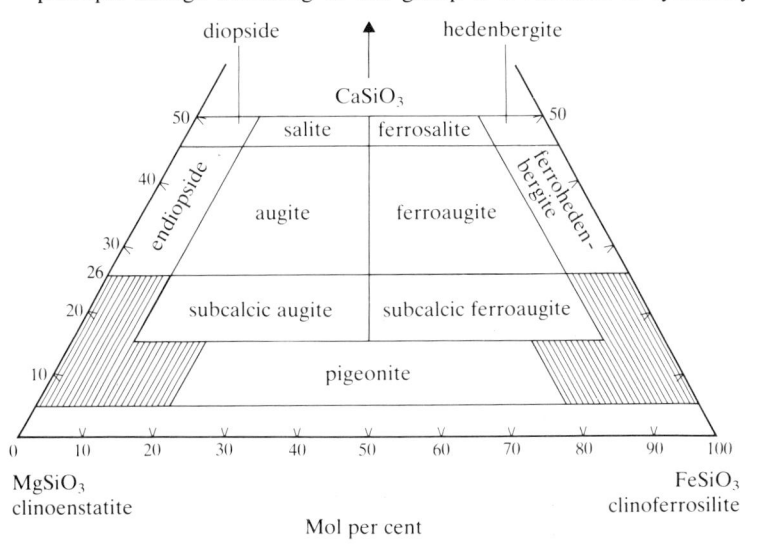

between the high Ca-bearing members (> 25% CaSiO$_3$), which are always monoclinic, and the low Ca-bearing members (< 15% CaSiO$_3$). In the latter group the iron-rich members (> 30% FeSiO$_3$) are monoclinic at high temperatures (pigeonites) but invert to an orthorhombic form (opx) at low temperatures. The magnesium-rich members (< 30% FeSiO$_3$) are orthorhombic at all temperatures. Orthopyroxenes can exist with up to 5 mol% CaSiO$_3$ in the structure.

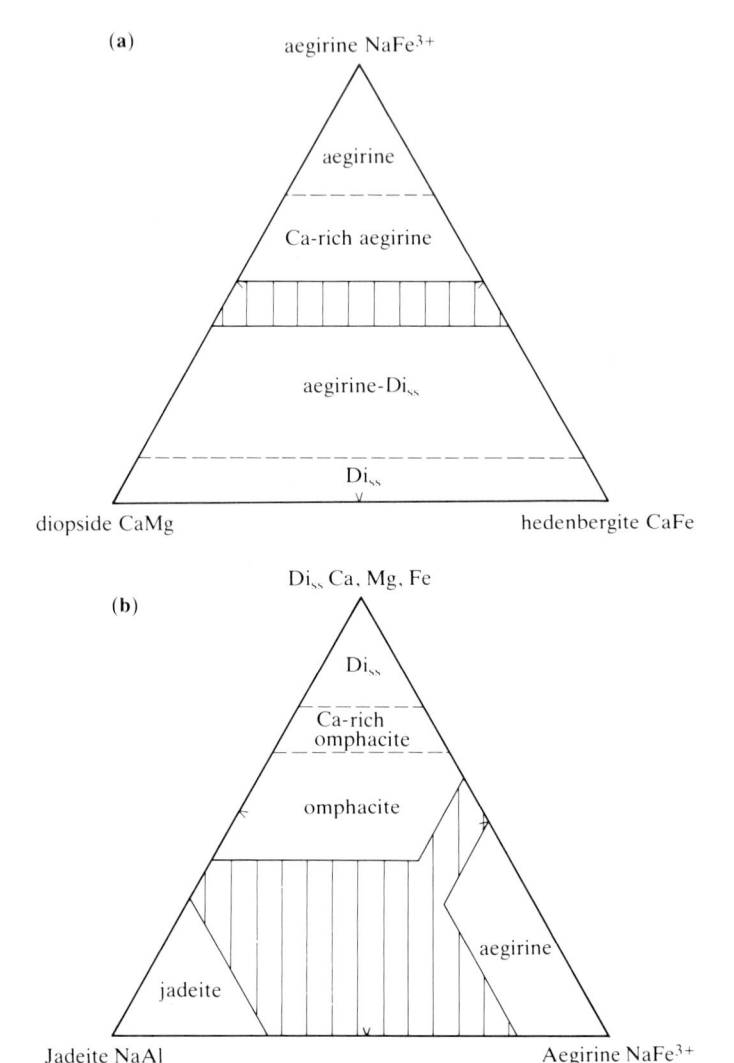

Figure 2.21 (a) Composition diagram for Na-pyroxenes (b) composition diagram for Na- and Al-pyroxenes.

The remaining pyroxenes of importance cannot be fitted into Figure 2.20 since they include sodium- and aluminium-bearing end members. Figure 2.21 shows two further systems to include the Na and Al members. In Figure 2.21a the central tie line from Figure 2.20, above $CaMgSiO_3$–$CaFeSiO_3$ (50 mol%) is used as the base of the triangle, and the other apex is $NaFe^{3+}SiO_3$. In Figure 2.21b the tie line $CaMgSiO_3$–$CaFe^{2+}SiO_3$, which represents diopside–hedenbergite or the diopside solid solution series (written as diopside$_{ss}$ or di$_{ss}$), now represents only one corner of the system, and the Na-bearing pyroxene $NaFe^{3+}SiO_3$ another. The third apex of the triangle is the phase $NaAlSiO_3$ or jadeite. The main mineral phases occurring are depicted in the two figures. The shaded area in Figure 2.21a means that a continuous sequence of change from di$_{ss}$ to aegirine ($NaFe^{3+}SiO_3$) does not exist. Similarly it will be observed from Figure 2.21b that the two main Al-bearing phases jadeite ($NaAlSiO_3$) and omphacite (a cpx containing Na, Al, Fe^{3+} and Mg, Fe^{2+}) are both isolated phases, quite separated from the other end member in the system.

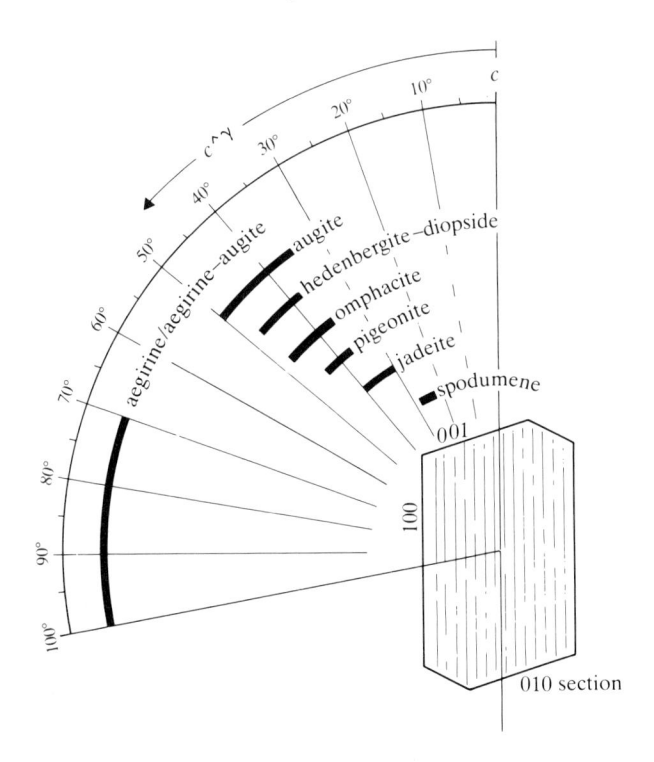

Figure 2.22 Extinction angles for clinopyroxenes. Ranges of $c\hat{}\gamma$ (maximum extinction angles) for several common pyroxene minerals. Note that $c\hat{}\alpha$ extinction angles for aegirine/aegirine–augite vary from 0° to ~ 20°.

SILICATE MINERALS

Figure 2.23 Exsolution lamellae parallel to (100).

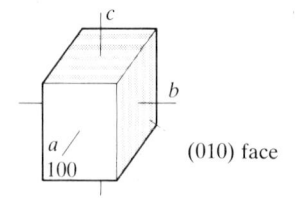

The 'normal' pyroxenes which are represented in Figure 2.20 are essential constituents of the basic calc-alkaline igneous rocks. These pyroxenes may occur in some ultrabasic and intermediate igneous rock types; and some pyroxenes in this system may occur in high temperature regional and thermal metamorphic rocks.

The Na-bearing pyroxenes primarily occur in alkaline igneous rocks of various types, and the Al-bearing pyroxenes occur in rocks (metamorphic or igneous) in which high pressures and temperatures have been operative.

The individual pyroxene minerals (or series) are discussed separately, but Figure 2.22 gives extinction angles for all the pyroxene minerals.

Exsolution lamellae

In many slowly-cooled pyroxenes, especially orthopyroxenes and augites, lamellae occur which have a definite crystallographic orientation. These are *not* twin lamellae but exsolved sheets with a different composition from the host mineral.

Orthopyroxenes may contain lamellae parallel to the (100) plane. These actually are Ca-rich clinopyroxene lamellae; the opx first crystallises at a high temperature with some calcium in the structure, then

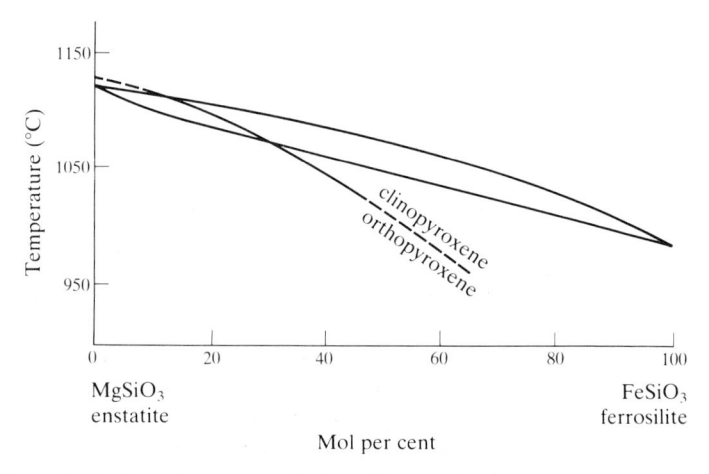

Figure 2.24 Inversion curve and phase diagram for pyroxenes.

Figure 2.25
Exsolution lamellae parallel to (001).

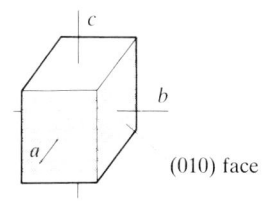

cools, and the excess calcium is exsolved as clinopyroxene lamellae parallel to the (100) plane (Fig. 2.23).

As crystallisation proceeds the ratio Mg:Fe in the liquid decreases. When the ratio reaches the value Mg:Fe = 70:30, orthopyroxene is replaced by a Ca-poor clinopyroxene (pigeonite). This Ca-poor clinopyroxene may also contain an excess of calcium at high temperature, which will be exsolved as Ca-rich clinopyroxene lamellae as the pigeonite cools. These exsolution lamellae are parallel to the (001) plane of the monoclinic pigeonite. As cooling proceeds, the Ca-depleted pigeonite cools through an 'inversion' curve, below which it changes (or inverts) to orthopyroxene (Fig. 2.24). The final result is a crystal of orthopyroxene containing lamellae of Ca-rich clinopyroxene parallel to the monoclinic (001) plane of the original pigeonite (Fig. 2.25).

Ca-rich clinopyroxenes also contain lamellae of exsolved pyroxene parallel to (100) and (001). The lamellae which are parallel to (100) are exsolved orthopyroxene, whereas the lamellae parallel to the (001) plane represent exsolved pigeonite. If cooling proceeds through the inversion curve, the pigeonite lamellae will invert to orthopyroxene, producing a second set of lamellae parallel to (100). Recent research work has shown that the crystallographic orientation of these lamellae, as well as their chemical composition, may be much more complex than originally was thought.

Crystallisation trends
Two Mg-rich pyroxenes may first crystallise from a basic magma, a Ca-poor orthopyroxene and a Ca-rich augite. As crystallisation proceeds, both pyroxenes become more Fe-rich until the Mg:Fe ratio reaches 70:30, at which point pigeonite replaces orthopyroxene as the Ca-poor pyroxene crystallising. Both clinopyroxenes become increasingly Fe-rich as crystallisation continues. In many intrusions crystallisation ceases at this point, but if fractionation is very marked, only one pyroxene (a Ca-rich ferroaugite) crystallises when the Mg:Fe ratio drops below 35:65. Finally, in extreme cases, crystallisation continues until there appears a Ca-rich ferrohedenbergite containing no Mg. Diagrammatically the crystallisation sequence, including olivine, can be depicted as follows:

SILICATE MINERALS

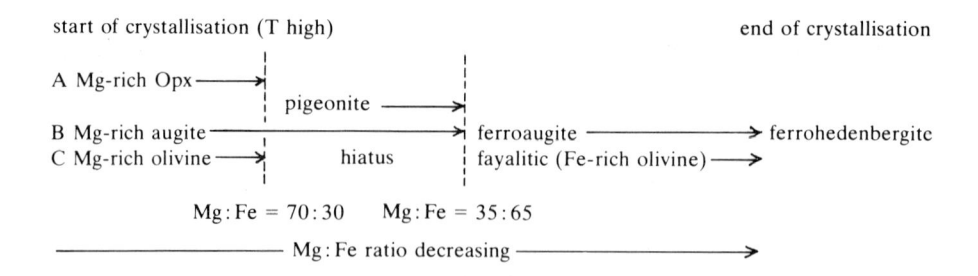

```
start of crystallisation (T high)                               end of crystallisation
A Mg-rich Opx ────▶
                    │ pigeonite ─────▶
B Mg-rich augite ───────────────▶│ ferroaugite ──────────▶ ferrohedenbergite
C Mg-rich olivine ──▶    hiatus   │ fayalitic (Fe-rich olivine) ──▶

        Mg:Fe = 70:30      Mg:Fe = 35:65
──────────────────── Mg:Fe ratio decreasing ────────────────────▶
```

It should be noted that the Mg:Fe ratios given above may change depending upon the amount of cations replacing Mg, Fe (and Ca) in the pyroxene structure. The above diagram can be represented as part of the cpx triangular diagram Figure 2.26: A, B and C are shown on the figure.

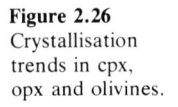

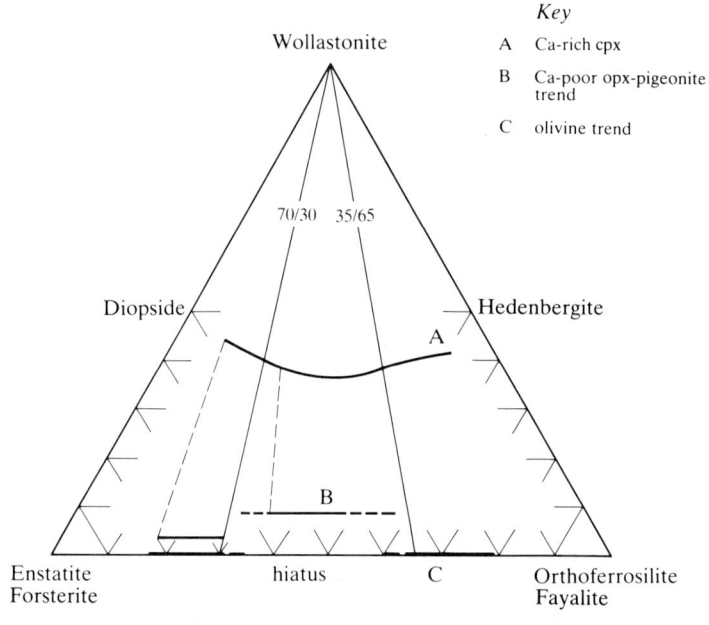

Figure 2.26
Crystallisation trends in cpx, opx and olivines.

Key
A Ca-rich cpx
B Ca-poor opx-pigeonite trend
C olivine trend

(Triangular diagram with vertices: Wollastonite (top), Enstatite/Forsterite (bottom left), Orthoferrosilite/Fayalite (bottom right); Diopside and Hedenbergite marked on sides; trend lines labelled 70/30 and 35/65; hiatus shown at base; curves labelled A, B, C.)

104

PYROXENE GROUP

Orthopyroxenes

Enstatite (En) – orthoferrosilite (Fs) $MgSiO_3$–$FeSiO_3$ orthorhombic
1.03 : 1 : 0.59 (En)

A complete sequence of orthopyroxenes is possible from 100% $MgSiO_3$ (100% enstatite) to 100% $FeSiO_3$ (100% orthoferrosilite). Pure enstatite can be written $En_{100}Fs_0$ and pure orthoferrosilite En_0Fs_{100}. Nowadays it is conventional to give the composition of an orthopyroxene as the percentage of orthoferrosilite present. Thus Fs_{24} means $En_{76}Fs_{24}$, or, to give its complete formula, $(Mg_{0.76}Fe_{0.24})SiO_3$.

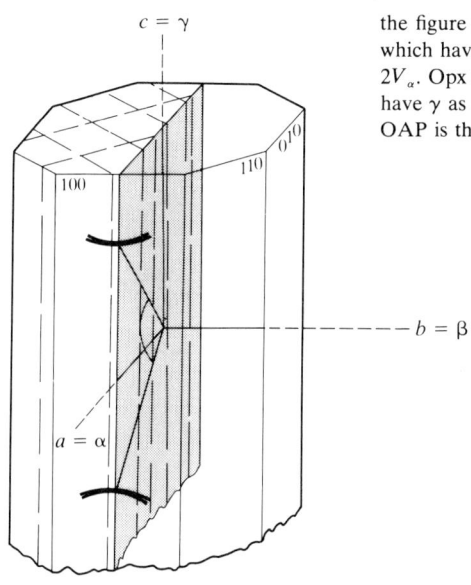

the figure shows opx Fs_{14} to Fs_{100} which have α as acute bisectrix $2V_\alpha$. Opx of composition Fs_0 to Fs_{13} have γ as acute bisectrix $2V_\gamma$ but OAP is the same

	En	Fs
n_α =	1.650–	1.768
n_β =	1.653–	1.770
n_γ =	1.658–	1.788
δ =	0.008–	0.020

$2V_\gamma$ = 60°–90° +ve (for opx of composition Fs_0–Fs_{13})
$2V_\alpha$ = 90°–50° −ve (for opx of composition Fs_{13}–Fs_{87})
High Fe-opx (Fs_{87}–Fs_{100}) are also positive with large $2V$ but these are not present in igneous rocks, and only occur in some meteorites. Figure 2.27 gives details of $2V$ variation.
OAP is parallel to (010). It is occasionally shown as being parallel to (100)
D = 3.21–3.96 (Fe-rich) H = 5–6

SILICATE MINERALS

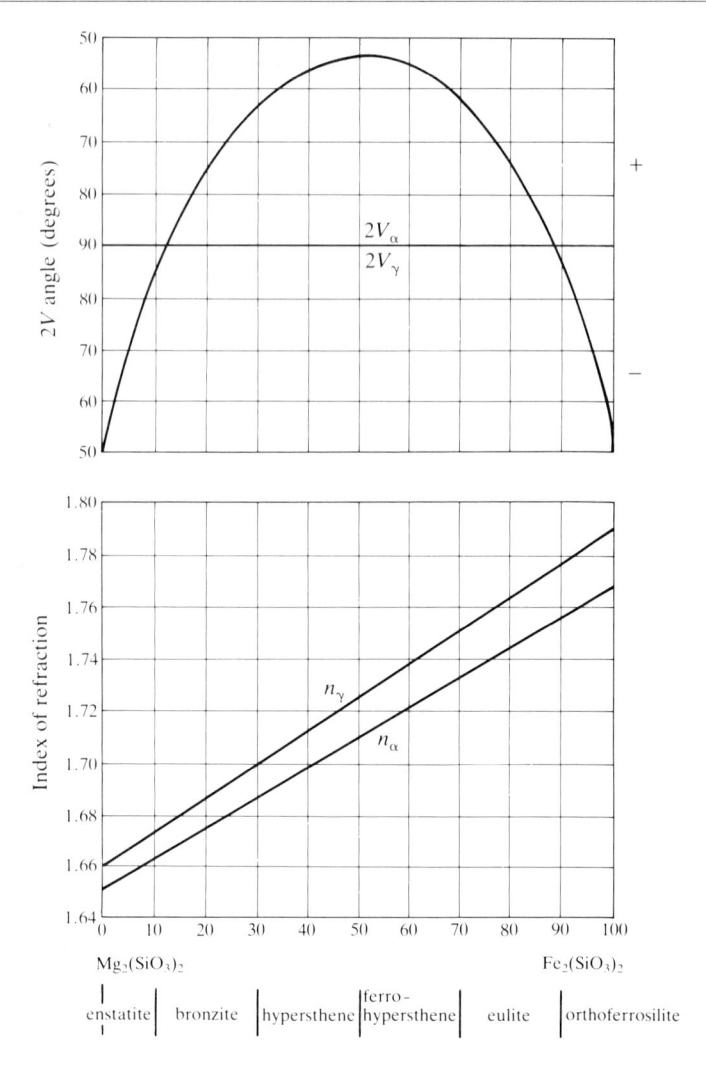

Figure 2.27 Variation of 2V angle and indices of refraction in the enstatite–orthoferrosilite (orthopyroxene) series.

COLOUR   Mg-rich opx are colourless; Fe-rich compositions show pale colours from pale green to pale brown.

PLEOCHROISM   Coloured opx show faint pleochroism with α pink to brown, β yellow to brown, and γ green, due to either Fe^{2+}, Ti or Al in the structure.

HABIT   Early formed opx in igneous rocks appear as short prismatic crystals.

*CLEAVAGE   Two good prismatic $\{110\}$ cleavages meet at nearly 90° on a basal section. $\{010\}$ and $\{100\}$ are poor cleavages or partings.

PYROXENE GROUP

RELIEF
: Moderate to high.

ALTERATION
: opx minerals alter to serpentine as follows:

$$3MgSiO_3 + 2H_2O \rightarrow Mg_3(Si_2O_5)(OH)_4 + SiO_2$$

The serpentine mineral is sometimes called bastite. Orthopyroxenes may occasionally alter to amphibole, cummingtonite first being formed:

$$8(Mg,Fe)SiO_3 + H_2O \rightarrow (Mg,Fe)_7Si_8O_{22}(OH)_2 + (Fe,Mg)O$$

A rim of amphibole is formed around the opx crystal and iron ores are released in the reaction, often seen in basic igneous plutonic rocks.

*BIREFRINGENCE
: Low with interference colours from low first order greys (En) to yellow and reds (iron-rich members).

INTERFERENCE FIGURE
: Large biaxial figures are seen on sections cut at right angles to the (100) plane + ve or − ve (see data above).

*EXTINCTION ANGLE
: All opx have straight extinction on prism edge or main prismatic cleavages.

TWINNING
: Absent from opx.

*OTHERS
: One set of exsolution lamellae is usually present, parallel to prismatic face (100). Another set may be present at a high angle to this. The explanation has been given under the heading 'Exsolution lamellae'.

DISTINGUISHING FEATURES
: Orthopyroxenes are distinguished from other clinopyroxenes by their parallel extinction. Opx is length slow whereas andalusite is length fast; sillimanite, although length slow, has a very small $2V$ and higher interference colours.

OCCURRENCE
: Orthopyroxenes occur in basic igneous rocks of all types. Mg-rich opx occurs in ultrabasic igneous rocks such as pyroxenites, harzburgites, lherzolites and picrites, in association with Mg olivine, Mg augite and Mg spinel. Orthopyroxenes occur in some regional metamorphic rocks, particularly charnockites and granulites, and may occur at high grades during the thermal metamorphism of argillaceous rocks in hornfelses of the innermost zones of thermal aureoles.

SILICATE MINERALS

Clinopyroxenes
Diopside (Di) – hedenbergite (Hed) $CaMgSi_2O_6$–$CaFeSi_2O_6$ monoclinic
Di 1.091 : 1 : 0.589, $\beta = 105°51'$
Hed 1.091 : 1 : 0.584, $\beta = 105°25'$

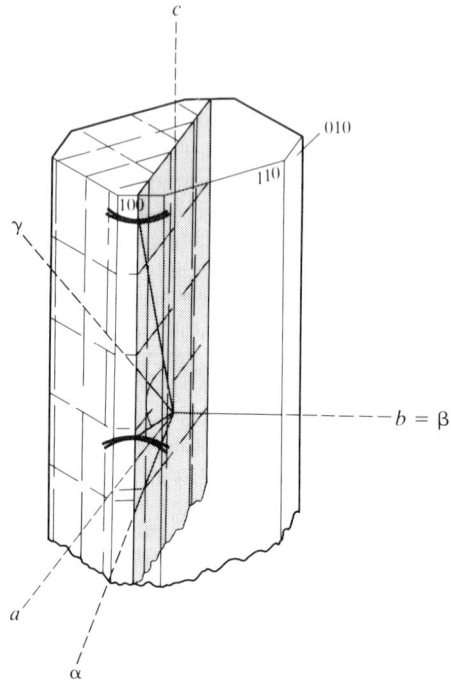

$n_\alpha = 1.664–1.726$
$n_\beta = 1.672–1.730$
$n_\gamma = 1.694–1.751$
$\delta = 0.030–0.025$
$2V_\gamma = 50°–62°$ +ve
OAP is parallel to (010)
$D = 3.22–3.56$ $H = 5\frac{1}{2}–6\frac{1}{2}$

COLOUR — Diopside is colourless and hedenbergite is brownish green.
PLEOCHROISM — Hedenbergites are pleochroic in pale greens and browns but pleochroism is weak and not a diagnostic feature.
HABIT — Occurs as short subhedral prisms.
*CLEAVAGE — {110} good prismatic cleavages meeting on basal section at 87°. Several partings {100}, {010} and {001} are present. A diopside with the {100} parting well developed is called diallage.
RELIEF — Moderate to high.
ALTERATION — Similar to that of the orthopyroxenes.
Diopsides will alter commonly to fibrous tremolite–actinolite, rarely to chlorite.

PYROXENE GROUP

*BIREFRINGENCE Moderate, middle second order greens and yellows common.
INTERFERENCE FIGURE Moderate $2V$ shown on a (100) section.

*EXTINCTION ANGLE Large extinction angle seen in an (010) section. The angle γ (slow ray)^cleavage is variable to over 45°. Extinction angles for all pyroxenes are shown in Figure 2.22.

TWINNING Simple and multiple twins common on $\{100\}$ and $\{001\}$.

OCCURRENCE Diopside occurs in a wide variety of metamorphic rocks, particularly metamorphosed dolomitic limestones and calcium-rich sediments. Diopside forms from the breakdown of the amphibole tremolite as the temperature increases. Diopside is usually accompanied by forsteritic olivine and calcite. Hedenbergite occurs in metamorphosed iron-rich sediments, being found in eulysites and skarns. Diopside may also occur in some basic extrusive igneous rocks, and hedenbergite may appear in some acid igneous rock types such as fayalite granite, fayalite ferro-gabbro, and some granophyres.

Pigeonite $Ca(Mg,Fe)Si_2O_6$ monoclinic
$\beta = 108°$

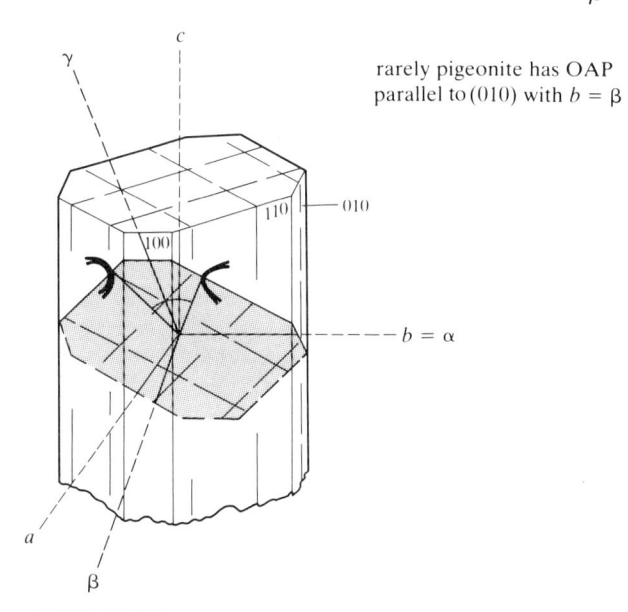

rarely pigeonite has OAP parallel to (010) with $b = \beta$

$n_\alpha = 1.682–1.722$
$n_\beta = 1.684–1.722$
$n_\gamma = 1.705–1.751$
$\delta = 0.023–0.029$
$2V_\gamma = 0–30°$ +ve
OAP is perpendicular to (010)
$D = 3.30–3.46$ $H = 6$

SILICATE MINERALS

OPTICAL PROPERTIES Very similar to diopside and augite except for $2V$ which is small ($< 30°$). A section cut parallel to face (101) will show a good Bx_a figure. The section is difficult to recognise but the interference colours will be very low (first order greys) and two cleavages will be present, meeting at an angle of less than 90°.

OCCURRENCE Pigeonite only occurs in rapidly chilled rocks. In most igneous rocks which have undergone slow cooling pigeonite is inverted to orthopyroxene.

Augite (ferroaugite) $Ca(Mg,Fe,Mn,Fe^{3+},Al,Ti)_2(Si,Al)_2O_6$ monoclinic
$$1.092:1:0.584, \beta = 105°50'$$

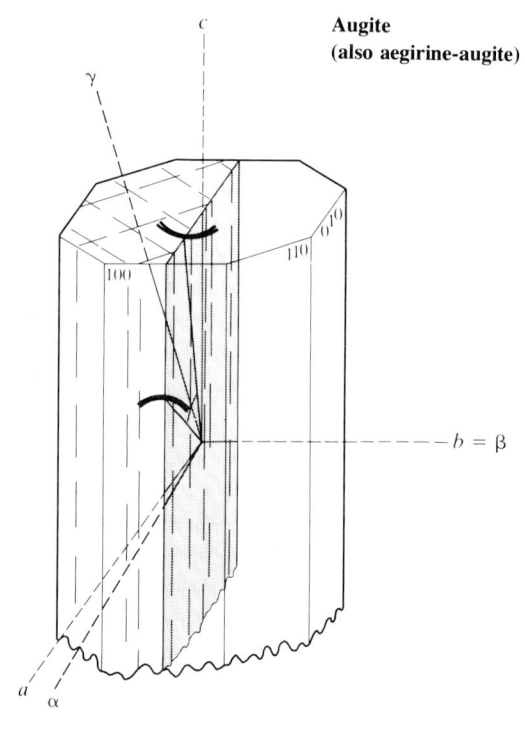

Augite
(also aegirine-augite)

$n_\alpha = 1.662–1.735$
$n_\beta = 1.670–1.741$
$n_\gamma = 1.688–1.761$
$\delta = 0.018–0.033$
$2V\gamma = 25°\ –55°$
$+\ ve$

Refractive indices change depending upon the Mg:Fe ratio, and also on the amount and lattice position of the minor constituents Al, Ti and Fe^{3+}. For example Al (Ti and Fe) in tetrahedral co-ordination (occupying Si sites) will increase $2V$ and lower RIs, whereas in the octahedral sites (occupying Mg and Fe positions) they will decrease $2V$ and increase RIs. These factors may affect compositions determined from optical properties by $\sim 5\%$

OAP is parallel to (010)
$D = 2.96–3.52$ $H = 5–6$

COLOUR	Augite is colourless to pale brown. Titanaugite (Ti-augite) is pale purple.
PLEOCHROISM	In most varieties pleochroism is very weak but titanaugite is weakly pleochroic with α pale green, β pale brownish and γ pale greenish purple.
HABIT	Variable from subhedral prismatic crystals in basic plutonic rocks to euhedral crystals in basic extrusive rocks.
*CLEAVAGE	Similar to diopside, with $\{110\}$ good and poor $\{100\}$ $\{010\}$ partings.
RELIEF	Moderate to high.
ALTERATION	Similar to diopside.
BIREFRINGENCE	Moderate, with maximum interference colours being low second order (blues, greens).
*INTERFERENCE FIGURE	Good Bx_a figure seen on the plane including the a and c axes, about the position of the face (101).
*EXTINCTION ANGLE	Similar to diopside, $\gamma\hat{}cl$ large (up to $45°+$).
TWINNING	Similar to diopside.
*OTHERS	Hourglass zoning is occasionally seen on certain prismatic sections especially in titanaugite.
DISTINGUISHING FEATURES	Augite is virtually indistinguishable from diopside. It may show a slighter smaller $2V$. Augite is found in mafic and ultramafic plutonic igneous rocks, whereas diopside occurs mostly in metamorphic rocks and basic volcanics.
OCCURRENCE	Augites occur mainly in igneous rocks and are essential mineral constituents of gabbros, dolerites and basalts. In plutonic gabbros augites frequently occur with orthopyroxenes (as already described under 'Crystallisation trends'). Augitic cpx have been recognised in some very high grade metamorphic rocks (granulites).

SILICATE MINERALS

Jadeite NaAlSi$_2$O$_6$

monoclinic
$1.103:1:0.613, \beta = 107°16'$

Jadeite

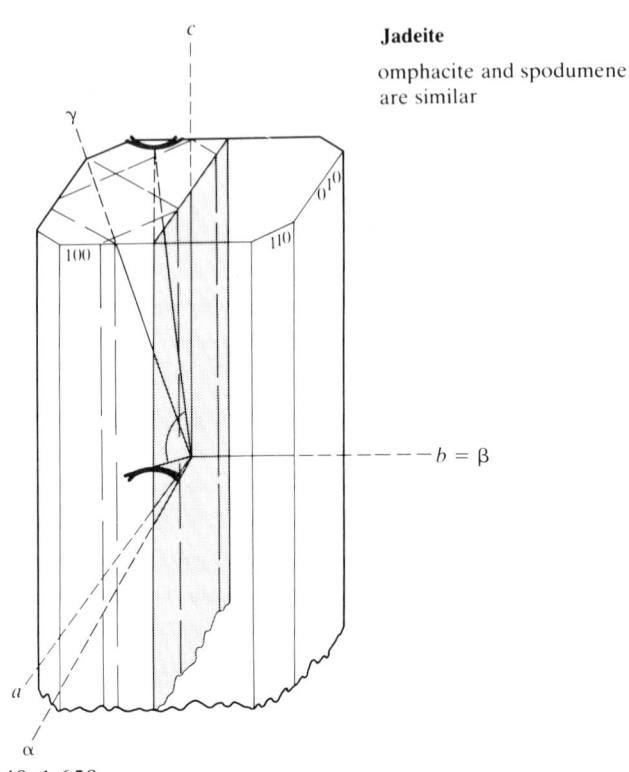

omphacite and spodumene are similar

$n_\alpha = 1.640–1.658$
$n_\beta = 1.645–1.663$
$n_\gamma = 1.652–1.673$
$\delta = 0.012–0.015$
$2V_\gamma = 67°–70°$ +ve
OAP is parallel to (010)
$D = 3.24–3.43$ $H = 6$

COLOUR Colourless.
HABIT Usually occurs as granular aggregates of crystals, usually medium to coarse grained.
CLEAVAGE Similar to diopside.
RELIEF Moderate.
ALTERATION Jadeite can alter to amphiboles or dissociate to a mixture of nepheline and albite.
BIREFRINGENCE Moderate with second order colours.
EXTINCTION ANGLE Similar to diopside, but smaller maximum angle with γ (slow)^cleavage = 33°–40° (max).

*OCCURRENCE Jadeite, is a rare pyroxene which can sometimes occur with albite in regional metamorphic rocks, especially glaucophane schists, which have formed under low heat flow, high P conditions. It is found in metamorphic rocks in association with lawsonite, glaucophane.

Omphacite is a cpx which has properties similar to jadeite and augite, and is found exclusively in eclogites. In this rock omphacite occurs with an Mg,Fe-garnet and the assemblage is considered to have formed at high T and P and be stable at temperatures below 700 °C with pressures of >10 kb. The higher the P of formation the richer the omphacite is in Na and Al.

SILICATE MINERALS

Aegirine (acmite) $NaFe^{3+}Si_2O_6$ $\qquad$ $1.099:1:0.601, \beta = 106°49'$
Aegirine–augite $(Na,Ca)(Fe,Fe^{3+},Mg)Si_2O_6$ $\qquad$ axial ratio variable

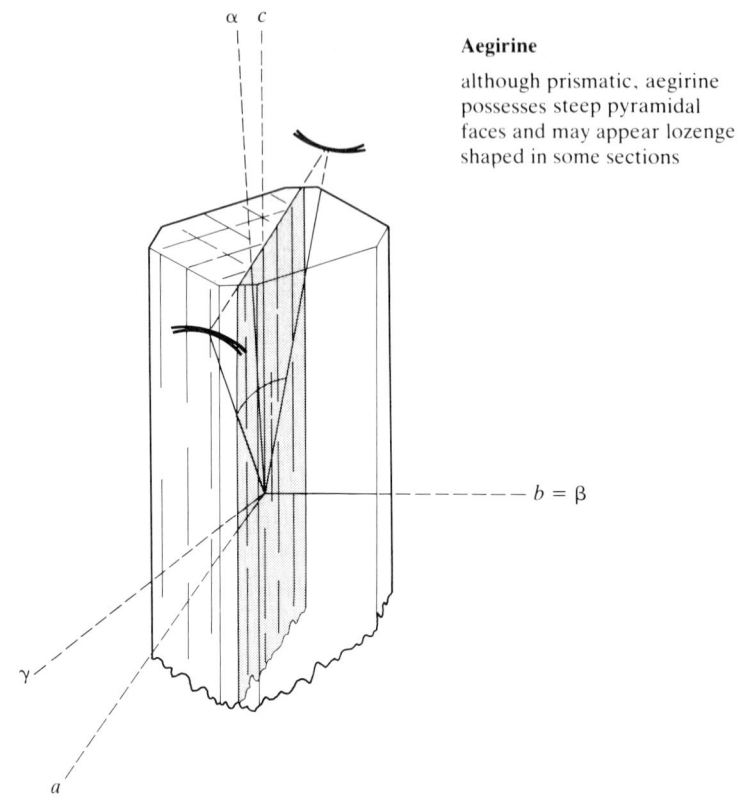

Aegirine

although prismatic, aegirine possesses steep pyramidal faces and may appear lozenge shaped in some sections

Aegirine	Aegirine–augite
$n_\alpha = 1.750$–1.776	$n_\alpha = 1.700$–1.750
$n_\beta = 1.780$–1.820	$n_\beta = 1.710$–1.780
$n_\gamma = 1.800$–1.836	$n_\gamma = 1.730$–1.800
$\delta = 0.040$–0.06	$\delta = 0.03$–0.05
$2V_\alpha = 40°$–$60°$ −ve	$2V_\alpha = 70°$–$90°$ −ve
OAP is parallel to (010)	$2V_\gamma = 90°$–$70°$ +ve
$D = 3.55$–3.60 $\quad H = 6$	OAP is parallel to (010)
	$D = 3.40$–3.55 $\quad H = 6$

*COLOUR Aegirine is strongly coloured in shades of green. Aegirine–augite is also coloured green in thin section.

*PLEOCHROISM Aegirine is strongly pleochroic, with α emerald green, β deep green and γ brownish green. Aegirine–augite is also pleochroic with a similar scheme to that of aegirine, but usually has a more yellowish colour throughout.

PYROXENE GROUP

HABIT Elongate prismatic crystals are most common.
CLEAVAGE Similar to other pyroxenes.
RELIEF High to very high.
*BIREFRINGENCE High but third order interference colours are usually masked by strong greenish colour which persists under crossed polars.
INTERFERENCE FIGURE $2V$ data given in Fig. 2.28.

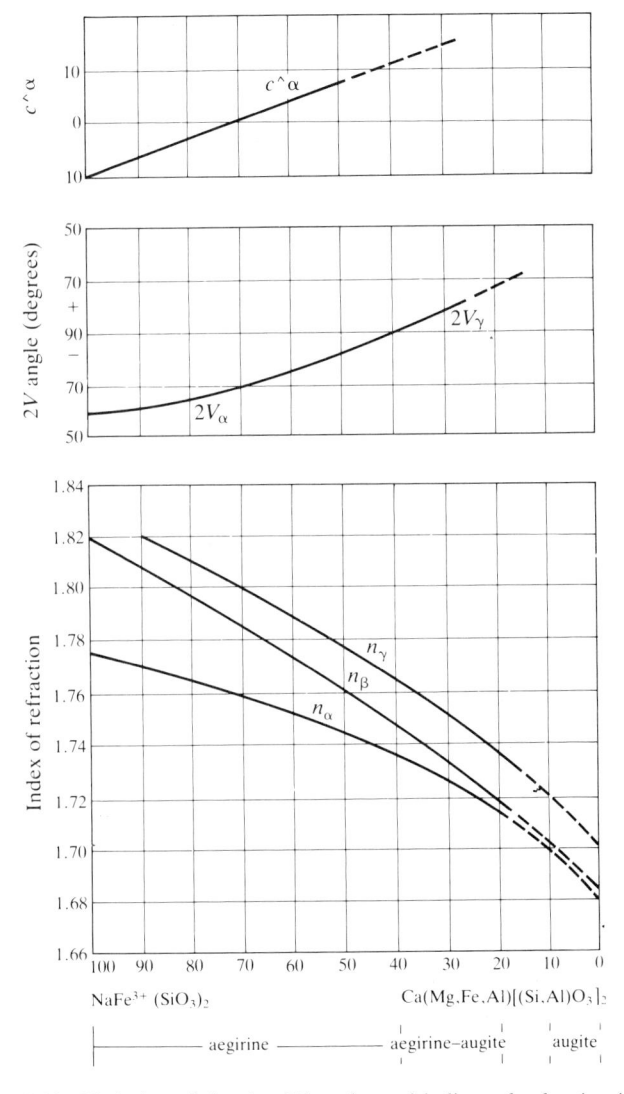

Figure 2.28 Variation of the $c\hat{\ }\alpha$, $2V$ angle, and indices of refraction in the aegirine–augite series (data from Deer, Howie & Zussman 1962).

SILICATE MINERALS

*EXTINCTION ANGLE Both aegirine and aegirine–augite have small extinction angles in an (010) prismatic section. The extinction angles α (fast)^c axis (or α^prismatic cleavage) = 0° to < 20°.

*OCCURRENCE Aegirine and aegirine–augite occur as late crystallisation products of alkali magmas, appearing in syenites and nepheline–syenites with alkali amphiboles (see earlier section in this chapter). They may occur in alkali granites, often with riebeckite, and may occur in some Na-rich schists with glaucophane and riebeckite.

Spodumene $LiAlSi_2O_6$ monoclinic
$1.144:1:0.632, \beta = 110°30'$

$n_\alpha = 1.648–1.663$
$n_\beta = 1.655–1.669$
$n_\gamma = 1.662–1.679$
$\delta = 0.014–0.027$
$2V_\gamma = 58°–68°$ +ve
OAP is parallel to (010)
$D = 3.03–3.22$ $H = 6½–7$

Properties of spodumene are similar to those of diopside – colourless in thin section and so on. But spodumene is a rare mineral, occurring in lithium-rich acid igneous rocks such as granite pegmatites, where it is associated with quartz, albite, lepidolite (lithium-rich mica), beryl and tourmaline.

Pseudo-pyroxenes

Wollastonite $CaSiO_3$ triclinic
$1.082:1:0.965$
$\alpha = 90°0', \beta = 95°16', \gamma = 103°22'$

Although used as an end member of the pyroxene group of minerals (Fig. 2.20) wollastonite does not possess a pyroxene structure, but is chemically similar and is described with them.

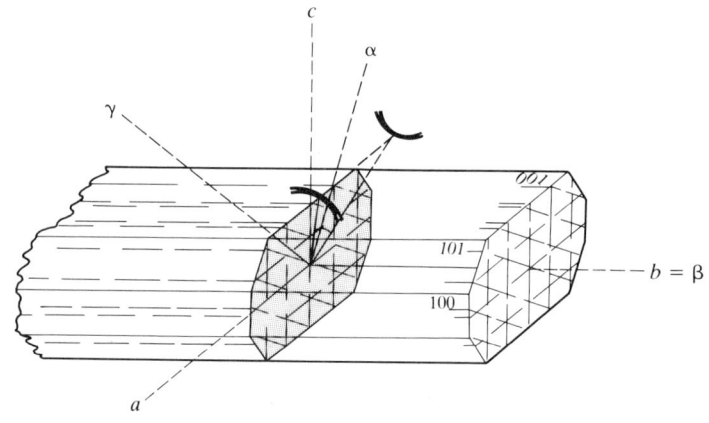

SCAPOLITE

n_α = 1.616–1.640
n_β = 1.628–1.650
n_γ = 1.631–1.650
δ = 0.013–0.014
$2V_\alpha$ = 38°–60° −ve
OAP is approximately parallel to (010)
D = 2.87–3.09 H = 4½–5

COLOUR — Colourless.
HABIT — Usually columnar or fibrous with rectangular cross sections.
*CLEAVAGE — $\{100\}$ perfect and $\{001\}$ and $\{10\bar{2}\}$ good. A typical section shows two or three cleavages.
RELIEF — Moderate.
BIREFRINGENCE — Low, with maximum first order yellow.
*INTERFERENCE FIGURE — Biaxial negative with a moderate $2V$, seen almost on the basal face (i.e. at right angles to the length of the crystals).
EXTINCTION ANGLE — Since the cleavages are different from those of pyroxene, the extinction angle is not so relevant a feature, but the crystals are almost length fast, and $\alpha\hat{\ }c$ axis (crystal length) is 30°–40°.
DISTINGUISHING FEATURES — Identification of wollastonite is difficult. However, it should be noted that although wollastonite is virtually identical to diopside, it is optically negative whereas diopside is optically positive.
OCCURRENCE — Wollastonite is a mineral formed in metamorphosed impure (calcareous) limestones, usually as a result of the reaction

$$CaCO_3 + SiO_2 \text{ (impurity)} \rightarrow CaSiO_3 + CO_2$$

at fairly high temperatures (about 1000 °C), which may be reduced if volatiles are present. Wollastonite has been recorded from some alkaline igneous rocks.

Parawollastonite (the monoclinic form of wollastonite) has a similar paragenesis.

Scapolite *Scapolite*   Tektosilicate

$(Ca,Na)_4[(Al,Si)_3Al_3Si_6O_{24}](Cl,CO_3)$ tetragonal c/a = 0.44

n_o = 1.540–1.600 ⎫ Indices increase with increasing substitution of Ca
n_e = 1.535–1.565 ⎭ for Na
δ = 0.004–0.037
Uniaxial −ve (length fast)
D = 2.50–2.78 H = 5–6

COLOUR — Colourless.
*HABIT — Large spongy prismatic crystals are common in metamorphosed carbonate rocks. Granular and fibrous-looking aggregates are also common, especially in garnet-bearing rocks.

SILICATE MINERALS

CLEAVAGE
: Good $\{100\}$ and $\{110\}$ prismatic cleavages.

RELIEF
: Low.

ALTERATION
: Scapolite may alter to a fine aggregate containing combinations of various minerals including chlorite, sercite, epidote, calcite, plagioclase, clays.

BIREFRINGENCE
: Low (Na-varieties) to moderate (Ca-varieties), with interference colours varying accordingly.

INTERFERENCE FIGURE
: Aggregates of crystals are usually too small for sign determination.

ZONING
: Compositional zoning is common.

DISTINGUISHING FEATURES
: Na-rich scapolite is similar in RI and birefringence to quartz, K-feldspar, plagioclase and cordierite, but it is uniaxial negative and untwinned whereas quartz is uniaxial positive. K-feldspars and cordierite are biaxial, and plagioclase is biaxial and invariably twinned. Nepheline is also uniaxial negative but its RIs are lower (often < CB) and it has a different occurrence.

OCCURRENCE
: Scapolite may occur in some pegmatites, replacing plagioclase or quartz, but its main occurrence is in metamorphic or metasomatic rocks.

Scapolite may form as a primary mineral in calcareous rocks subjected to regional metamorphism at amphibolite facies. It is associated with calcite, sphene, diopside, plagioclase, epidote and garnet. In contact metamorphism scapolite forms in carbonate rocks by introduction of sodium and chlorine from the igneous intrusion. Grossular garnet, wollastonite, and fluorite are associated minerals.

Serpentine

Serpentine $Mg_3Si_2O_5(OH)_4$

Phyllosilicate

monoclinic

$0.57:1:1.31, \beta = 93°$

Serpentine includes a variety of minerals, one fibrous (chrysotile) and two tabular (lizardite and antigorite)

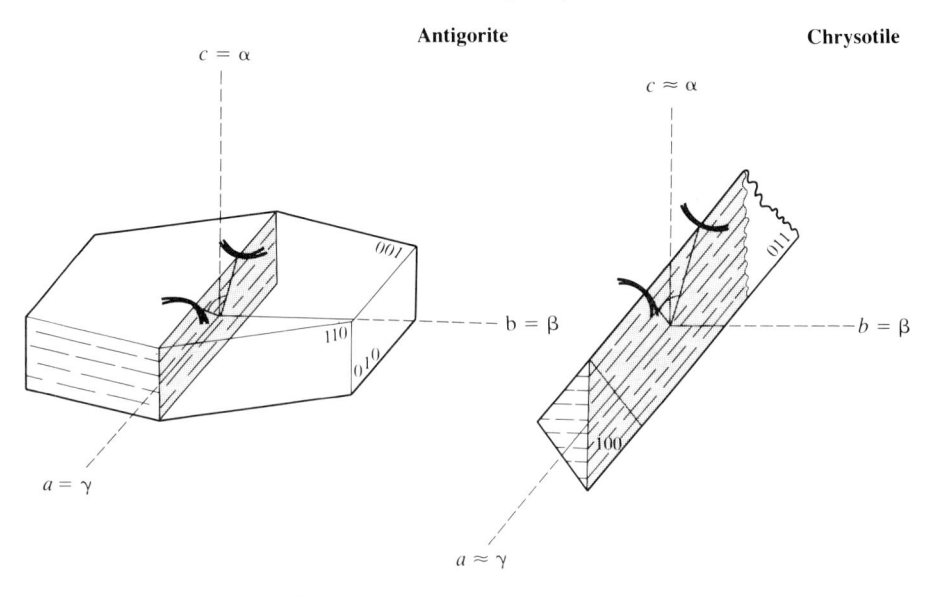

	Chrysotile	Lizardite	Antigorite
n_α	1.53–1.55	1.54–1.55	1.56–1.57
n_β	–	–	1.57
n_γ	1.55–1.56	1.55–1.56	1.56–1.57
δ	0.013–0.017	0.006–0.008	0.004–0.007
$2V_\alpha$	variable –ve	?	37°–61° –ve
OAP	parallel to (010)	?	parallel to (010)
D	2.55	2.55	2.6
H	2½	2½	2–3½

COLOUR Colourless to pale green.
HABIT Chrysotile is fibrous elongated parallel to the *a* crystallographic axis, and lizardite and antigorite are both flat, tabular crystals.
CLEAVAGE Chrysotile has a fibrous cleavage, and lizardite a basal cleavage.
RELIEF Low.
*BIREFRINGENCE Low or very low, often with anomalous pale yellow colours shown (cf. chlorite).
INTERFERENCE FIGURE Antigorite shows a medium sized negative $2V$ on a basal section.
EXTINCTION Straight on fibres, cleavage or crystal edge. Chrysotile is length slow.

SILICATE MINERALS

DISTINGUISHING FEATURES
Serpentine minerals have lower birefringence and lower refractive indices than chlorite and fibrous amphiboles. Most chlorites exhibit either stronger birefringence or anomalous interference colours. Brucite can show anomalous colours similar to chlorite, but brucite is uniaxial.

*OCCURRENCE
Serpentine minerals are formed during the alteration of ultrabasic igneous rocks – dunites, pyroxenites and peridotites – at temperatures below 400 °C. Chrysotile probably forms first and antigorite then is derived from it.

Chrysotile is the major variety of commercial asbestos and occurs as economic deposits in Canada, South Africa and Russia.

Silica group   Tektosilicates

The various forms of silica (SiO_2 is the formula for all silica minerals) can be represented on a simple $P-T$ diagram (Fig. 2.29). This shows that the lowest temperature form of quartz, called α-quartz (or low quartz), inverts to β-quartz (or high quartz) at 573 °C at atmospheric pressure; the temperature of this inversion increases with increasing pressure ($\sim$ 670 °C at 3 kb). At 867 °C, β-quartz inverts to tridymite; the temperature of this inversion also increases considerably with increasing pressure (to $\sim$ 1450 °C at 3 kb). Tridymite inverts to cristobalite at

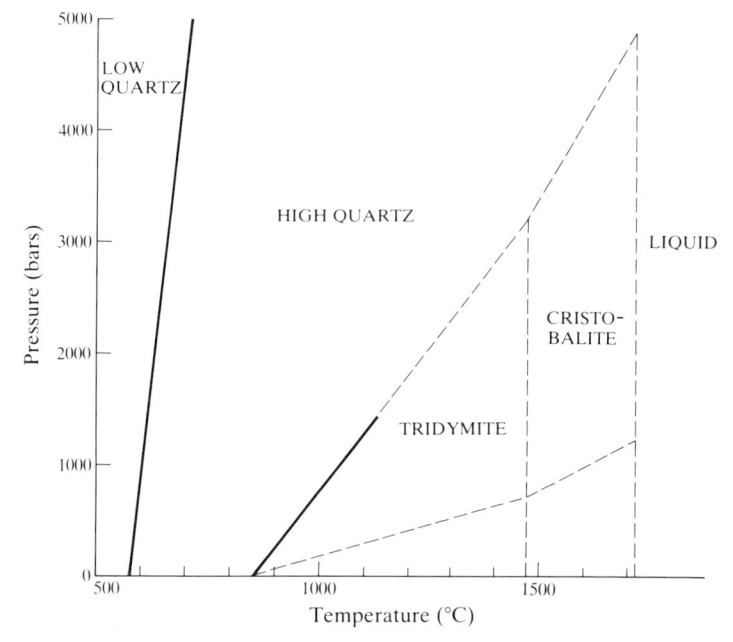

Figure 2.29 Pressure–temperature diagram for SiO_2 (after Tuttle & Bowen 1958).

SILICA GROUP

1470 °C at atmospheric pressure, and the temperature of this inversion does not change with increasing pressure. Finally, at 1713 °C cristobalite melts and the liquidus boundary is reached. This diagram can, of course, be interpreted in the other direction, with liquid SiO_2 crystallising; it can then be determined which polymorph (minerals with same chemistry but different structure) will be encountered at which temperature as crystallisation proceeds.

Other forms of silica, not included in the diagram are coesite, a high pressure phase formed at high temperature (400–800 °C) with stishovite, a high density collapsed structural form found at very high pressures and temperatures. It is found in meteorite impact craters, and minerals possessing stishovite structures may exist in the Earth's upper mantle. Quartz is an essential constituent in acid igneous rocks and arenaceous sedimentary rocks, and is a common constituent in metamorphic rocks. The three main quartz minerals are described.

Quartz SiO_2 trigonal, c/a 1.09997

$n_o = 1.544$
$n_e = 1.553$
$\delta = 0.009$
Uniaxial +ve (length slow)
$D = 2.65$ $H = 7$

*COLOUR   Colourless.

HABIT   Euhedral quartz crystals are prisms with hexagonal cross sections, and may appear as phenocrysts in acid extrusive rocks, but quartz usually occurs as shapeless interstitial grains in igneous and metamorphic rocks or as rounded grains in sedimentary clastic rocks.

*CLEAVAGE   None.

RELIEF   Low, just greater than 1.54.

*ALTERATION   None.

*BIREFRINGENCE   Low, maximum interference colours are first order white or pale yellow.

EXTINCTION   Straight on prism edge.

TWINNING   Numerous types of twins occur, in particular Brazil (twin plane 11$\bar{2}$0) and Dauphiné (twin axis is c axis), but twinning is not detectable under the microscope because optic orientation in both twin parts is identical in both types of twin.

OTHERS   In some porphyritic acid extrusive and hypabyssal igneous rocks where quartz occurs as phenocrysts, the crystals may show corroded margins because of a reaction between the quartz and the magmatic liquid.

OCCURRENCE   Quartz is an essential mineral in acid igneous plutonic rocks such as granites and granodiorites, but may be present in diorites and some gabbros. In these, quartz occurs as shapeless grains. In rapidly cooled extrusive and hypabyssal rocks, for example rhyolites, dacites, pitchstones and quartz porphyries, quartz may occur as euhedral phenocrysts. It is also found as large late formed crystals in pegmatites, and is a

SILICATE MINERALS

common constituent in hydrothermal veins accompanying various ore minerals. Quartz is one of the most common detrital minerals owing to its lack of cleavage, hardness and stability. Because of this quartz is a common and often essential mineral in coarse terrigeneous rocks such as conglomerates and sandstones, and also occurs in siltstones and mudstones where its fine grain size is such that detection may not be possible with a microscope. In many sedimentary rocks, including some limestones, authigenic quartz will form. The secondary quartz actually grows around pre-existing grains or forms well-developed crystals, and is formed during diagenesis after deposition of the sediments. Quartz occurs in many metamorphic rocks, usually remaining in the rocks until high grades of metamorphism are reached. At these highest grades a reaction

$$\text{muscovite} + \text{quartz} \rightarrow \text{K-feldspar} + \text{sillimanite}$$

may occur, and the grain size of these high grade gneisses will be about the same as granites (greater than 3 mm).

Tridymite SiO_2 orthorhombic

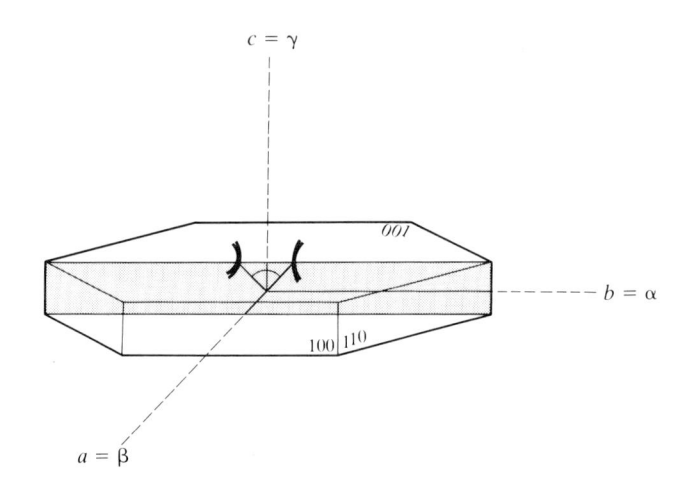

n_α = 1.469–1.479
n_β = 1.470–1.480
n_γ = 1.473–1.483
δ = 0.004
$2V_\gamma$ = 40°–90° +ve
OAP is parallel to (100)
D = 2.26 H = 7

COLOUR Colourless.
*HABIT Six sided tabular crystals common, but usually very tiny.

SILICA GROUP

CLEAVAGE A poor prismatic cleavage occurs.
RELIEF Moderate, RI considerably less than 1.54.
ALTERATION None.
BIREFRINGENCE Low: very low first order colours.
INTERFERENCE FIGURE Very difficult to obtain because of very small crystal size.
*TWINNING Common on $\{110\}$, seen as wedge shaped (sector type) twinning on basal plane.
OCCURRENCE Rare in rocks although tridymite may be found in quickly cooled igneous rocks such as rhyolites, pitchstones, dacites and so on. It may be found in association with sanidine (see under 'Feldspar group') and sometimes augite or fayalitic olivine. Tridymite has been recorded from high temperature thermally metamorphosed impure limestones.

Cristobalite SiO_2 tetragonal, c/a 1.395

$n_o = 1.478$
$n_e = 1.484$
$\delta = 0.003$
Uniaxial $-$ve
$D = 2.38 \quad H = 6\text{--}7$

COLOUR Colourless.
*HABIT Minute square sectioned crystals are common, but often occurs as skeletal fibrous crystals in cavities.
CLEAVAGE None.
RELIEF Moderate, considerably less than 1.54.
ALTERATION None.
BIREFRINGENCE Very weak.
INTERFERENCE FIGURE Basal sections give a uniaxial negative figure but minute crystal size makes this difficult to obtain.
TWINNING Twins on $\{111\}$ but not seen in thin section.
OCCURRENCE Usually found in cavities in volcanic rocks, and has been discovered in some thermally metamorphosed sandstones. Since both tridymite and cristobalite occur as metastable forms outside their stability field, no conclusions can be drawn about the P–T conditions of formation of rocks containing these minerals.

SILICATE MINERALS

Sphene   Nesosilicate

Sphene $CaTiSiO_4(O,OH,F)$ monoclinic
$0.755:1:0.854, \beta = 119°43'$

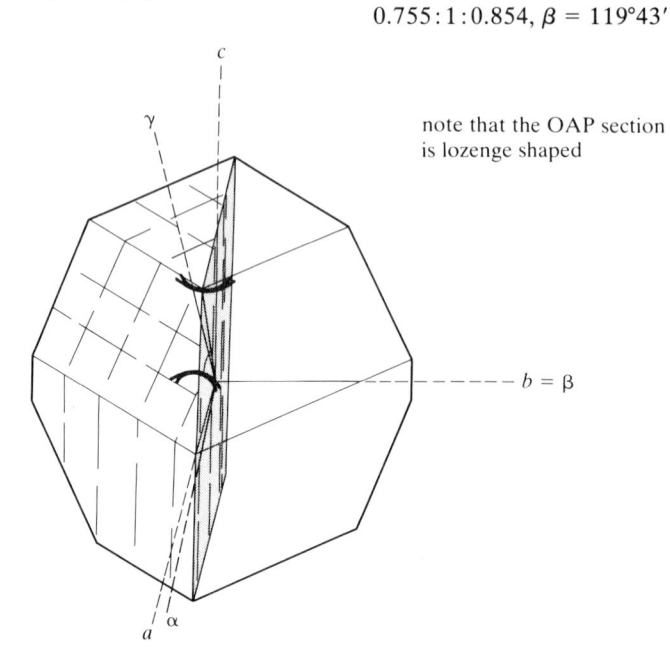

note that the OAP section is lozenge shaped

$n_\alpha = 1.843–1.950$
$n_\beta = 1.879–2.034$
$n_\gamma = 1.943–2.110$
$\delta = 0.100–0.192$
$2V_\gamma = 17°–40°$ (usually $23°–34°$) +ve
OAP is parallel to (010)
$D = 3.45$ $H = 5$

COLOUR   Colourless, pale brown, dark brown.
PLEOCHROISM   Well displayed in sphenes from alkali igneous plutonic rocks, with α yellow or colourless, β pinkish, yellow brown, γ pink, yellow, orange brown.
*HABIT   Anhedral to euhedral, often occurring as small lozenge or diamond shaped crystals.
*CLEAVAGE   {110} good.
*RELIEF   Extremely high.
ALTERATION   Sphene alters to leucoxene, an aggregate of quartz and rutile to which ilmenite also alters, as follows:

$$2CaTiSiO_4(O,OH,F) \rightarrow \underbrace{2TiO_2 + 2SiO_2}_{\text{leucoxene}} + 2Ca(O,OH)$$

STAUROLITE

*BIREFRINGENCE Extreme, but colours tend to be masked by body colour and high relief of mineral. (This mineral does not change appearance under crossed polars.)

INTERFERENCE FIGURE $2V$ is small; the small size of most crystals, coupled with the high dispersion of light, means that a good figure is not often obtained.

*TWINNING Single twins with $\{100\}$ twin plane common.

OCCURRENCE It is a common accessory mineral in igneous and metamorphic rocks, particularly plutonic intermediate and acid rocks, such as diorite, granodiorites and granites and also alkali igneous rocks such as nepheline–syenites. It is common in some skarns and in metamorphic basic schists and gneisses. It is rare in sediments.

Staurolite

Nesosilicate

Staurolite $(Fe,Mg)_2(Al,Fe^{3+})_9O_6Si_4O_{16}(O,OH)_2$

orthorhombic
$0.473:1:0.341$

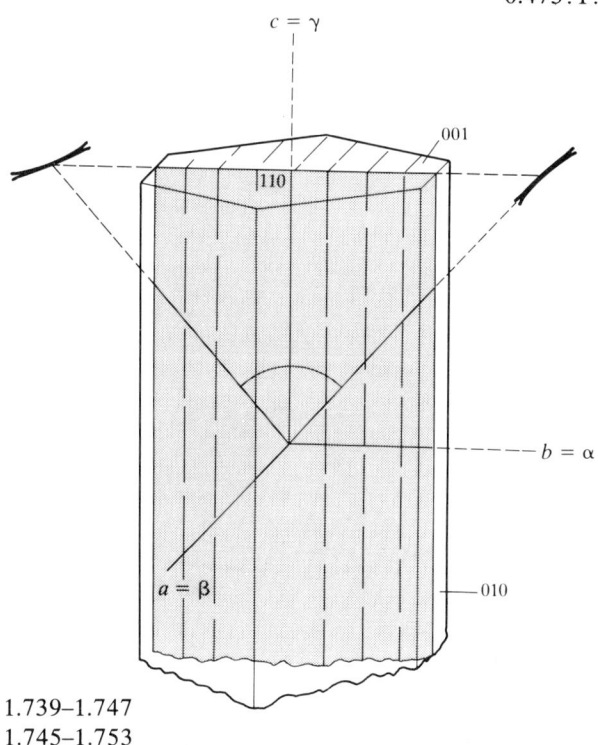

$n_\alpha = 1.739–1.747$
$n_\beta = 1.745–1.753$
$n_\gamma = 1.752–1.761$
$\delta = 0.013–0.014$
$2V_\gamma = 82°–90°$ +ve
OAP is parallel to (100)
$D = 3.74–3.85$ $H = 7\frac{1}{2}$

SILICATE MINERALS

*COLOUR	Yellow, pale yellow.
*PLEOCHROISM	Always present and distinct in yellows with α colourless, β pale yellow, and γ rich golden yellow.
HABIT	Staurolite occurs as squat prisms, usually containing inclusions, particularly of quartz.
CLEAVAGE	$\{010\}$ moderate.
RELIEF	High.
*ALTERATION	Rare, but may alter to a green ferric chlorite.
BIREFRINGENCE	Low, but interference colours are masked by the yellow colour of mineral.
INTERFERENCE FIGURE	Since $2V$ is very large, a single optic axis is required.
EXTINCTION	Straight on prism edge or cleavage.
OCCURRENCE	Staurolite occurs only in regional metamorphic rocks which are rich in alumina and iron, with probably a high $Fe^{3+}:Fe^{2+}$ ratio. In medium grade pelitic rocks, staurolite develops from chloritoid and before kyanite but kyanite and staurolite can coexist. With increasing grade staurolite breaks down to give kyanite and garnet.

Talc   Phyllosilicate

Talc $Mg_6Si_8O_{20}(OH)_4$ monoclinic
$0.577:1:2.068, \beta = 98°55'$

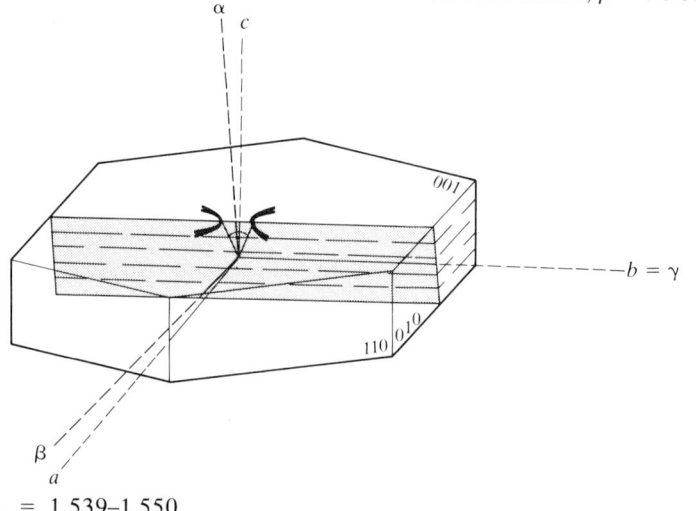

$n_\alpha = 1.539–1.550$
$n_\beta = 1.589–1.594$
$n_\gamma = 1.589–1.600$
$\delta = 0.05$
$2V_\alpha = 0°–30°$ $-ve$
OAP is perpendicular to (010)
$D = 2.58–2.83$ $H = 1$

TOPAZ

COLOUR Colourless.
HABIT Tabular crystals or pseudo-hexagonal plates, similar to micas.
*CLEAVAGE Perfect {001} basal.
RELIEF Low.
*BIREFRINGENCE High with interference colours of third order. Basal sections, similar to those of muscovite, give very low first order greys.
INTERFERENCE FIGURE Good Bx_a figure with small $2V$ on basal section.
EXTINCTION Straight.
*OCCURRENCE Talc occurs by low grade thermal metamorphism of siliceous dolomites, and by the hydrothermal alteration of ultrabasic rocks, where talc may occur along faults and shear planes. The development of talc is often associated with serpentinisation, with the serpentine changing to talc plus magnesite by addition of CO_2.

Topaz Nesosilicate

Topaz $Al_2SiO_4(OH,F)_2$ orthorhombic, $0.528:1:0.955$

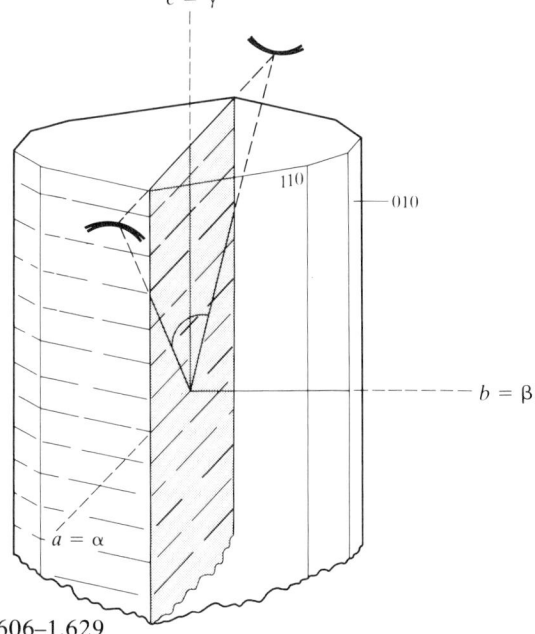

$n_\alpha = 1.606–1.629$
$n_\beta = 1.609–1.631$
$n_\gamma = 1.616–1.638$
$\delta = 0.008–0.011$
$2V_\gamma = 48°–68°$ +ve
OAP is parallel to (010) prism section is length slow
$D = 3.49–3.57$ $H = 8$

SILICATE MINERALS

COLOUR Colourless but thick sections may be yellowish or pink.
HABIT Usually occurs as prismatic crystals subhedral to anhedral.
*CLEAVAGE {001} perfect.
RELIEF Moderate.
*INCLUSIONS Liquid inclusions are common either of water or brewsterlinite (originally thought to be CO_2).
*ALTERATION Topaz alters to clay minerals (kaolin) and sericite. One reaction is as follows:

$$2Al_2SiO_4(OH,F)_2 + 2H_2O + 2SiO_2 \rightarrow Al_4Si_4O_{10}(OH,F)_8$$
$$\text{topaz} \hspace{6cm} \text{kaolin}$$

This is a hydrothermal alteration taking place at a late stage, in the presence of free silica.

BIREFRINGENCE Low.
INTERFERENCE FIGURE A single optic axis figure is required to obtain sign.
*OCCURRENCE Topaz is a mineral found in late stage acid igneous rocks such as granites, rhyolites and pegmatites, where it can also occur in cavities. It is associated with pneumatolytic action and is a constituent of greisen. It occurs with quartz, cassiterite, tourmaline, fluorite and beryl. It has been known to form in metamorphosed bauxite deposits.

Tourmaline Cyclosilicate

Tourmaline $Na(Mg,Fe,Mn,Li,Al)_3Al_6Si_6O_{18}(BO_3)_3(OH,F)_4$ trigonal, c/a 0.447

The composition of tourmaline varies greatly, with many types known. Thus dravite is the magnesian variety ($NaMg_3$ etc.), schorl the ferromanganoan variety ($Na(Fe,Mn)_3$ etc.) and elbaite the lithium-bearing type ($Na(Li,Al)_3$ etc.).

$n_o = 1.610–1.630+$
$n_e = 1.635–1.655+$ RI depends on composition
$\delta = 0.021–0.026+$

Uniaxial −ve (all types) (prismatic sections are length fast)
$D = 2.9–3.2$ $H = 7–7\frac{1}{2}$

*COLOUR Highly variable; colourless, blue, green, yellow.
*PLEOCHROISM Elbaite is usually colourless, but the other varieties are pleochroic:

dravite	o	dark brown	e	pale yellow
	o	yellow brown	e	yellow
schorl	o	dark green	e	reddish violet
	o	blue	e	pale green, pale yellow

*HABIT Tourmaline almost always occurs as large elongate prismatic crystals, often occurring in radiating clusters.

CLEAVAGE $\{11\bar{2}0\}$ and $\{10\bar{1}1\}$ poor, often appearing perpendicular to the prism zone.
RELIEF Moderate.
BIREFRINGENCE Moderate, colours of second order seen but frequently masked if tourmaline has a strong body colour.
TWINNING Rare.
ZONING Colour zoning may occur.
OCCURRENCE Tourmaline typically occurs in granite pegmatites, pneumatolytic veins and some granites as the schorl–elbaite type. In the pneumatolytic stage of alteration, tourmaline may form by boron introduction, and the rock luxullianite forms in this environment. In pneumatolytic igneous assemblages tourmaline is associated with topaz, spodumene, cassiterite, fluorite and apatite. In metamorphic rocks (especially metamorphosed limestones) and metasomatic rocks, the dravite type of tourmaline occurs; and dravites have been recorded from basic igneous rocks. Tourmalines have been found as detrital minerals in sedimentary rocks, and as authigenic minerals in some limestones.

Vesuvianite — Sorosilicate

Vesuvianite (or idiocrase) $Ca_{10}(Mg,Fe)_2Al_4Si_9O_{34}(OH,F)_4$ tetragonal, c/a 0.757

n_o = 1.708–1.752
n_e = 1.700–1.746
δ = 0.001–0.008
Uniaxial $-$ve (prism section is length fast)
D = 3.33–3.43 H = 6–7

COLOUR Colourless, pale yellow, pale brown.
HABIT Prismatic crystals usually occur, but in general crystals are subhedral with only a few faces present.
CLEAVAGE $\{110\}$ and $\{100\}$ poor.
RELIEF High.
BIREFRINGENCE Low, greys of first order.

Vesuvianite is a difficult mineral to recognise; in relief and birefringence it resembles zoisite (see 'Epidote group'). However, it usually occurs as large mineral grains and its occurrence and mineral associations are most important.

*OCCURRENCE Vesuvianite or idiocrase occurs in thermally metamorphosed limestones and in skarns. It is associated with grossular (Ca-bearing) garnet, diopside and wollastonite. It has been found in nepheline–syenites and in veins in basic igneous rocks.

Zeolite group — Tektosilicates

The zeolites are hydrated alumino silicates of K, Na and Ca. They occur in amygdales and vesicles in basic extrusive rocks where they are

SILICATE MINERALS

deposited by late stage hydrothermal solutions. Analcime is closely related to this group of minerals, which also includes natrolite ($Na_2Al_2Si_3O_{10}.2H_2O$). The zeolites are widely used as indicator minerals in thick lava piles, such as ocean floor basalts, to determine temperature and depth of burial. A typical sequence from a recent Icelandic lava pile is:

top
| zeolite-free zone
| chabazite–thomsonite $CaAl_2Si_4O_{12}.6H_2O$ –
| $[NaCa_2(Al,Si)_5O_{10}]_2.6H_2O$
| analcime (+ natrolite) $NaAlSi_2O_6.H_2O$ (for natrolite
| see above)
| mesolite–scolecite $[Na_2Ca_2Al_2Si_3O_{10}]_3.8H_2O$ –
bottom $CaAl_2Si_3O_{10}.3H_2O$

Although other areas of extrusive rocks may show slight variations in the zeolites present, the zones described above generally occur. Natrolite and most other zeolites are colourless in thin section, with RIs very much lower than the cement. They mostly belong to orthorhombic or monoclinic crystal systems (natrolite is orthorhombic), with either straight extinction or small extinction angle. Natrolite is length slow. Their $2V$ is usually large +ve or −ve and their birefringence is variable but low. Their occurrence in vesicles and amygdales is the most reliable indicator for identification. X-ray diffraction techniques are required for positive identification of zeolite type. The main optical properties of the zeolites are as follows:

*COLOUR	Colourless.
HABIT	Apart from analcime (see 'Feldspathoid family'), most zeolites are elongate fibrous or platy, often occupying cavities or amygdales in extrusive igneous rocks.
CLEAVAGE	Variable depending upon crystal system. Most fibrous varieties possess at least one prismatic cleavage.
*RELIEF	Low to moderate; RI is less than 1.54 for all minerals.
ALTERATION	Rare, but a few zeolites will alter to clay minerals.
*BIREFRINGENCE	Generally low to very low. A very few zeolites may show first order yellow.
INTERFERENCE FIGURE	Variable.
EXTINCTION	All fibrous varieties have straight extinction on prism edge except for scolecite. Platy varieties usually possess inclined extinction.
TWINNING	Simple twinning is common in mesolite, laumonite, chabazite and stilbite. Multiple twinning is common in scolecite, phillipsite and harmotome.

Zircon

Zircon ZrSiO$_4$

Nesosilicate

tetragonal, c/a 0.891

n_o = 1.923–1.960
n_e = 1.968–2.015
δ = 0.042–0.065
Uniaxial +ve (prism section is length slow)
D = 4.6–4.7 H = 7½

COLOUR Colourless pale brown.
*HABIT Very small, squat, square prisms occur with terminal faces. Zircons are usually found as euhedral crystals.
CLEAVAGE $\{110\}$ imperfect; $\{111\}$ poor.
*RELIEF Extremely high.
ALTERATION None.
*BIREFRINGENCE Very high, a prismatic crystal section will show third or fourth order interference colours.
TWINNING Rare.
ZONING May be present due to outer metamict zones on an unaltered core.
DISTINGUISHING FEATURES Tiny euhedral crystals in alkaline or acid plutonic igneous rocks are usually zircon. Sphene is pale brown with usually a diamond-shaped cross section and is biaxial +ve. Monazite is biaxial +ve. Cassiterite and rutile are coloured minerals.
*OCCURRENCE An accessory mineral found in all igneous rocks, but essentially in intermediate to acid varieties, where it is associated with biotite crystals. Haloes frequently occur in the biotite surrounding minute zircon crystals (due to radioactive decay of U and Th damaging the biotite structure by β particle bombardment). Zircon is most commonly found in plutonic igneous rocks, particularly granites, granodiorites, diorites, syenites, nepheline–syenites and pegmatite veins. Zircon also occurs as a detrital mineral in sediments, and will survive through many metamorphic and melting events.

3 The non-silicates

3.1 Introduction

Minerals which are not silicates have been grouped together in this chapter for the description of their properties. However, unlike the silicates, the crystal structures and chemical variation of members of the group are not easily related to mineralogical properties and mode of occurrence. Even subdivision of the group into transparent and opaque minerals is impractical, as closely related minerals and even compositional varieties of the same mineral may vary in opacity. For example, sphalerite is transparent when it is pure zinc sulphide but becomes more opaque with increasing iron substitution of zinc.

The non-silicates can usually be regarded as accessory minerals in most rocks, yet they are major components in some rock types, e.g. halides in evaporites, sulphides in massive sulphide deposits and carbonates in limestones.

Minerals of the following non-silicate groups appear in this chapter: carbonates (CO_3^{2-}), sulphides (S^{2-}), oxides (O^{2-}), halides (Cl^-, F^-), hydroxides (OH^-), sulphates (SO_4^{2-}), phosphates (PO_4^{4-}), tungstates (WO_4^{2-}) and native elements. Within each group the minerals are described in alphabetical order. The relationship of optical and physical properties to chemical composition and structure is outlined only for the first four groups.

In this chapter, where appropriate, thin-section information is as described in Section 1.3 and presented for the silicates in Chapter 2. The polished-section information, using reflected light, is as described in Section 1.6.

3.2 Carbonates

The carbonates, of which the most well known example is calcite $CaCO_3$, contain a discrete $(CO_3)^{2-}$ radical that may be considered as a single anion in the structure but is in fact a trigonal planar complex. This complex, with carbon in the centre of an equilateral triangle formed by three oxygens, is shown in the carbonate structure in Figure 3.1. There are relatively few common carbonates of rock-forming significance, and most can be considered as secondary or replacive minerals forming on alteration of metal-bearing precursor minerals, e.g. cerussite $PbCO_3$ after galena PbS. Some secondary carbonates contain structural water, e.g. malachite $Cu_2CO_3(OH)_2$ after chalcopyrite $CuFeS_2$.

CARBONATES

Figure 3.1 The structure of calcite $CaCO_3$.

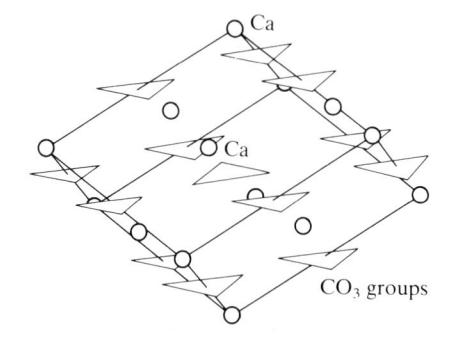

CO₃ groups

The triangular nature of the $(CO_3)^{2-}$ radical dominates the structure of the carbonates and results in trigonal (rhombohedral) or orthorhombic (pseudo-hexagonal) symmetry. The critical factor controlling the type of symmetry is the radius of the dominant metallic cation; for elements such as Mn, Fe, Mg with radius less than about 1.0 Å the carbonates are trigonal, but for elements such as Ba, Sr, Pb with larger radii the carbonates are orthorhombic. Calcium lies close in radius value to the critical size, and this explains the existence of $CaCO_3$ as two minerals, calcite (trigonal) and aragonite (orthorhomic). Although aragonite is considered to be a high pressure polymorph of $CaCO_3$, it can grow at low pressures provided that the solution chemistry is correct. However, it is metastable and usually inverts to calcite during recrystallisation processes such as diagenesis.

Figure 3.2 Carbonates in the $CaCO_3$–$MgCO_3$–$FeCO_3$ system.

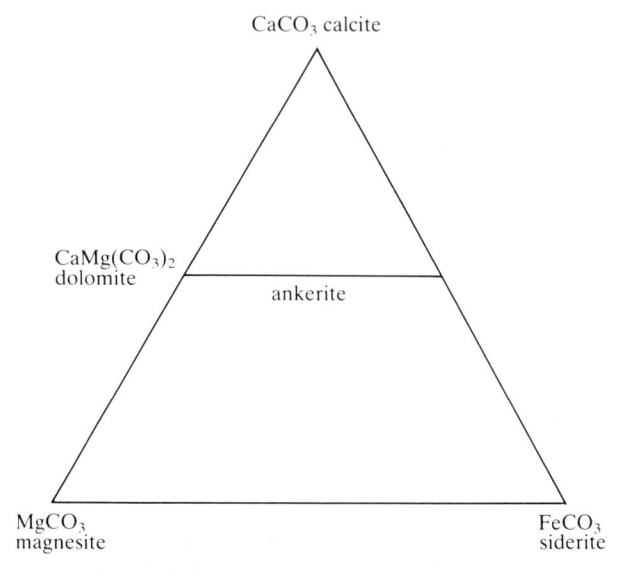

$CaCO_3$ calcite

$CaMg(CO_3)_2$ dolomite

ankerite

$MgCO_3$ magnesite

$FeCO_3$ siderite

THE NON-SILICATES

Table 3.1 Optical properties of the common carbonates.

Trigonal structures (uniaxial)	n_o	n_e	Optic sign
calcite $CaCO_3$	1.658	1.486	$-ve$
dolomite $CaMg(CO_3)_2$	1.679	1.500	$-ve$
siderite $FeCO_3$	1.875	1.635	$-ve$
rhodochrosite $MnCO_3$	1.816	1.597	$-ve$

Orthorhombic structures (biaxial)	n_α	n_β	n_γ	$2V$	Optic sign
strontianite $SrCO_3$	1.518	1.665	1.667	8°	$-ve$
witherite $BaCO_3$	1.529	1.676	1.677	16°	$-ve$
aragonite $CaCO_3$	1.530	1.680	1.685	18°	$-ve$

Chemical substitution is quite significant in the common carbonates, e.g. manganoan calcite $(Ca,Mn)CO_3$ and magnesian siderite $(Fe,Mg)CO_3$, but substitution is probably most extensive in dolomite (see Fig. 3.2, the triangular diagram $CaCO_3$–$MgCO_3$–$FeCO_3$).

Minerals of the carbonate group have very large birefringences (Table 3.1) and they usually have well developed cleavages and multiple twinning. The large birefringence is due to the planar triangular $(CO_3)^{2-}$ radicals which are orientated normal to the c axis. That a mineral is a carbonate is usually easily determined in thin section. However, identification of the particular carbonate usually requires selective chemical staining or chemical analysis. In polished section, the carbonates have low reflectance values but have distinct bireflectance due to the large birefringence. Identification of the particular carbonate is again difficult, but it is useful to remember that reflectance depends on refractive index; calcite and dolomite in intergrowths are usually much more readily distinguished on the basis of reflectance in polished section than on the basis of relief in thin section.

Alteration, due to oxidation, of iron-bearing carbonates leads to a penetrative yellowish or reddish brown staining, whereas manganese-bearing carbonates yield a black alteration product.

Aragonite $CaCO_3$ orthorhombic
     $0.6228:1:0.7204$

$n_\alpha = 1.530$
$n_\beta = 1.680$
$n_\gamma = 1.685$
$\delta = 0.155$
$2V_\alpha = 18°$ $-ve$ (crystals are length fast)
OAP is parallel to (100)
$D = 2.94$ $H = 3\frac{1}{2}$

CARBONATES

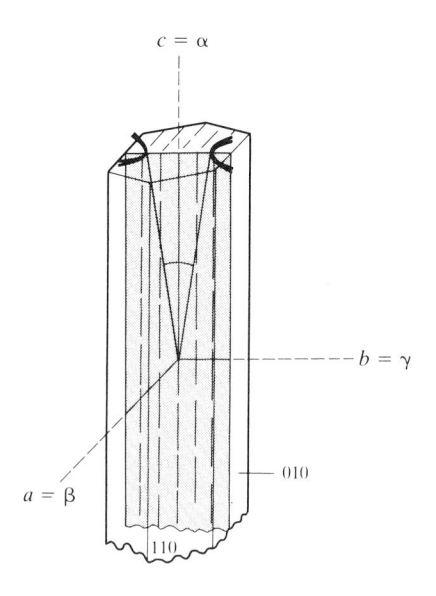

- COLOUR — Colourless.
- *HABIT — Thin prismatic or occasionally fibrous crystals occur as for example in shell structures.
- CLEAVAGE — {010} prismatic cleavage imperfect.
  Low to moderate but variable with optic orientation, as for calcite.
- RELIEF — Minimum RI is parallel to c axis (i.e. parallel to prism length).
- *BIREFRINGENCE — Extremely high, similar to calcite.
- INTERFERENCE FIGURE — Difficult to obtain because of crystal size, but good Bx_a figure may be seen on basal section ($2V$ very small).
- EXTINCTION — Straight on cleavage or prism edge.
- TWINNING — Common, lamellar twins on {110}, parallel to c axis. Repeated twinning also common.
- OCCURRENCE — Aragonite is less common than calcite. Many invertebrates build their shells of aragonite, which gradually changes to calcite on diagensis. Thus pre-Mesozoic fossil shells will inevitably consist of calcite. Aragonite occurs as a secondary mineral, often in association with zeolites, in cavities in volcanic rocks. It is a widespread metamorphic mineral in glaucophane–schist facies metamorphic rocks in which deep burial produces aragonite as the stable carbonate at ~ 300 °C and 6 to 10 kb pressure. Aragonite inversion to calcite may occur as the rock attains normal P and T conditions.

Calcite $CaCO_3$ trigonal, c/a 0.8543

n_o = 1.658
n_e = 1.486
δ = 0.172
Uniaxial $-$ve
D = 2.715 H = 3

COLOUR Colourless.

HABIT Often as shapeless grains (anhedral); occasional rhombohedral outline seen in sedimentary limestones.

CLEAVAGE Perfect $\{10\bar{1}0\}$ rhombohedral cleavage – three cleavage traces seen in some sections.

*RELIEF Moderate with extreme variation because of large birefringence. Note that the refractive indices cover a range of values which 'bracket' 1.54. The crystal is said to 'twinkle' during rotation. Prismatic crystals parallel to the c axis are length fast (n_e).

*BIREFRINGENCE ⎫
INTERFERENCE ⎬ Extremely high with pale pinks and greens of fourth order and higher. Because of the large birefringence, grains show moderate order interference colours even when the optic axis is near vertical, and these can be used to obtain a uniaxial interference figure.
FIGURE ⎭

TWINNING $\{01\bar{1}2\}$ common, appearing as multiple twins, $\{0001\}$ common, simple twin plane.

OCCURRENCE One of the most common non-silicate minerals. It is a principal constituent of sedimentary limestones, occurring as carbonate shell material, as fine precipitates, and as clastic material. Shells generally are composed of calcite or aragonite. Aragonite usually occurs as the initial carbonate precipitate but it eventually recrystallises to calcite.

On metamorphism pure calcitic limestone recrystallises to marble in which the calcite grains are welded together in a mosaic; in impure limestones the calcite combines with impurites present to give new minerals, the type of mineral depending upon the temperature and CO_2 pressure. The reaction

$$\text{calcite} + \text{silica} \rightarrow \text{wollastonite (CaSiO}_3\text{)} + CO_2$$

occurs at $\approx$ 600 °C at low pressures but the same reaction occurs at over 800 °C as the pressure increases. Calcite can also occur with calc-silicate minerals such as diopside, garnet (Ca-rich) and idocrase (vesuvianite) in metamorphic rocks.

Calcite may occur in vugs or cavities in igneous rocks, being deposited by late stage hydrothermal solutions. In hydrothermal veins, calcite is a common gangue mineral, often being found with fluorite, barite or quartz and in association with the sulphide ore minerals. Calcite may occur as a primary crystallising mineral in some igneous rocks, particularly carbonatites and some nepheline–syenites. Calcite may also occur as a secondary mineral on alteration of ferromagnesian minerals by late stage hydrothermal solutions in which CO_2 is present.

Dolomite $CaMg(CO_3)_2$ trigonal, c/a 0.80

$n_o = 1.679$
$n_e = 1.500$
$\delta = 0.179$
Uniaxial $-$ve
$D = 2.86 \quad H = 3\frac{1}{2}–4$

COLOUR — Colourless.
*HABIT — Usually subhedral, but dolomitisation of limestones often leads to euhedral crystals occurring as rhombohedra with curved faces (baroque dolomite).
CLEAVAGE — Perfect $\{10\bar{1}1\}$ rhombohedral, as calcite.
RELIEF — Low to moderate (variable with optic orientation).
*BIREFRINGENCE — Extremely high, even higher than calcite (almost colourless but slight iridescence gives indication of extreme birefringence).
*TWINNING — Similar to calcite, i.e. multiple on $\{02\bar{2}1\}$. The twin lamellae show birefringence of a lower order than the crystal.
ZONING — Commonly encountered, Fe^{2+} substitution of Mg^{2+} (ankerite).
OCCURRENCE — Note that dolomite is also a name given to rock consisting mainly of dolomite. Dolomite occasionally occurs as a primary mineral in sedimentary rocks and is often associated with evaporite deposits. As a secondary mineral dolomite is formed during dolomitisation of a limestone shortly after deposition, and before consolidation. Another type of dolomitisation occurs after consolidation of a limestone if Mg-rich solutions enter the rock. Dolomite is currently forming in certain saline lakes. The formation of primary and secondary dolomite may be due to the marine environment changing from deep to shallow water with increasing salinity. Dolomite can occur as a gangue mineral with fluorite, barite, calcite, quartz or siderite in association with lead and zinc sulphides. Dolomite rock is commonly associated with serpentines and other ultramafic rocks and it is common in ophiolite suites. During metamorphism dolomitic marbles may crystallise from dolomitic limestones. At higher grades of metamorphism the dolomite eventually breaks down to give periclase MgO, with brucite, $Mg(OH)_2$ forming on hydration.

Siderite $FeCO_3$ trigonal, c/a 0.8184

$n_o = 1.782$
$n_e = 1.575$
$\delta = 0.207$
Uniaxial $-$ve
$D = 3.50 \quad H = 4+$

COLOUR Colourless to pale brown or pale yellow.
HABIT Euhedral crystals are common, and siderite is often found as aggregates of crystals in oolitic structures.
All other properties similar to calcite (note extreme birefringence).
OCCURRENCE Common in ironstone nodules in Carboniferous argillaceous rocks and also in the Jurassic ironstones of central England. In Raasay in the Inner Hebrides siderite is associated with chamosite.

Siderite is found in veins with other gangue minerals and metallic ores.

3.3 Sulphides

In the structures of sulphide minerals, sulphur atoms are usually surrounded by metallic atoms (e.g. Cu, Zn, Fe) or the semi-metals (Sb, As or Bi). The chemical bonding is usually considered to be essentially covalent. Although sulphur has a preference for fourfold tetrahedral co-ordination it is found in a large variety of co-ordination polyhedra which may be quite asymmetric. Non-stoichiometry, i.e. a variable metal:sulphur ratio, is a feature of many sulphide structures, especially at high temperatures; complex ordering may result on cooling of a non-stoichiometric phase leading, at low temperature, to minerals with only slightly different compositions but different structures. A good example is that of high temperature cubic digenite, $Cu_{2-x}S$ ($x \leq 0.2$), which is represented at low temperatures by orthorhombic chalcocite Cu_2S, orthorhombic djurleite $Cu_{1.97}S$ and cubic digenite $Cu_{1.8}S$.

Two further possible complexities in sulphide structures are the existence of sulphur–sulphur bonds exemplified by the S_2^{2-} pair in pyrite FeS_2 (see Fig. 3.3), and the existence of structures that can be considered as resulting from a replacement by a semi-metal of half the sulphur in such pairs, e.g. arsenopyrite FeAsS.

Most sulphides are opaque but some (e.g. sphalerite when pure zinc sulphide) are transparent. Some are transparent for red light (e.g. pyragyrite Ag_3SbS_3) or only in the infra-red (e.g. stibnite Sb_2S_3). Many are semiconductors, which means that they conduct electricity at a high temperature but not at a low temperature. In fact, the optical and physical properties of many sulphides are best understood if the band model of semiconductors is applied (see Shuey 1975).

The structures of several common sulphides are illustrated in Figure 3.3. As is evident from the few examples given, sulphide structures can be classified – as are the silicates – into structures based on chains, sheets, networks and so on. Although such a classification is of less value than for the silicates, consideration of structures in such a way helps to explain crystal morphology, cleavage directions etc. of some sulphides.

The sulphosalts are one group of sulphides which are very diverse chemically and structurally. They contain a semi-metal as well as a metal

Figure 3.3 Sulphide structures (after Vaughan & Craig 1978).

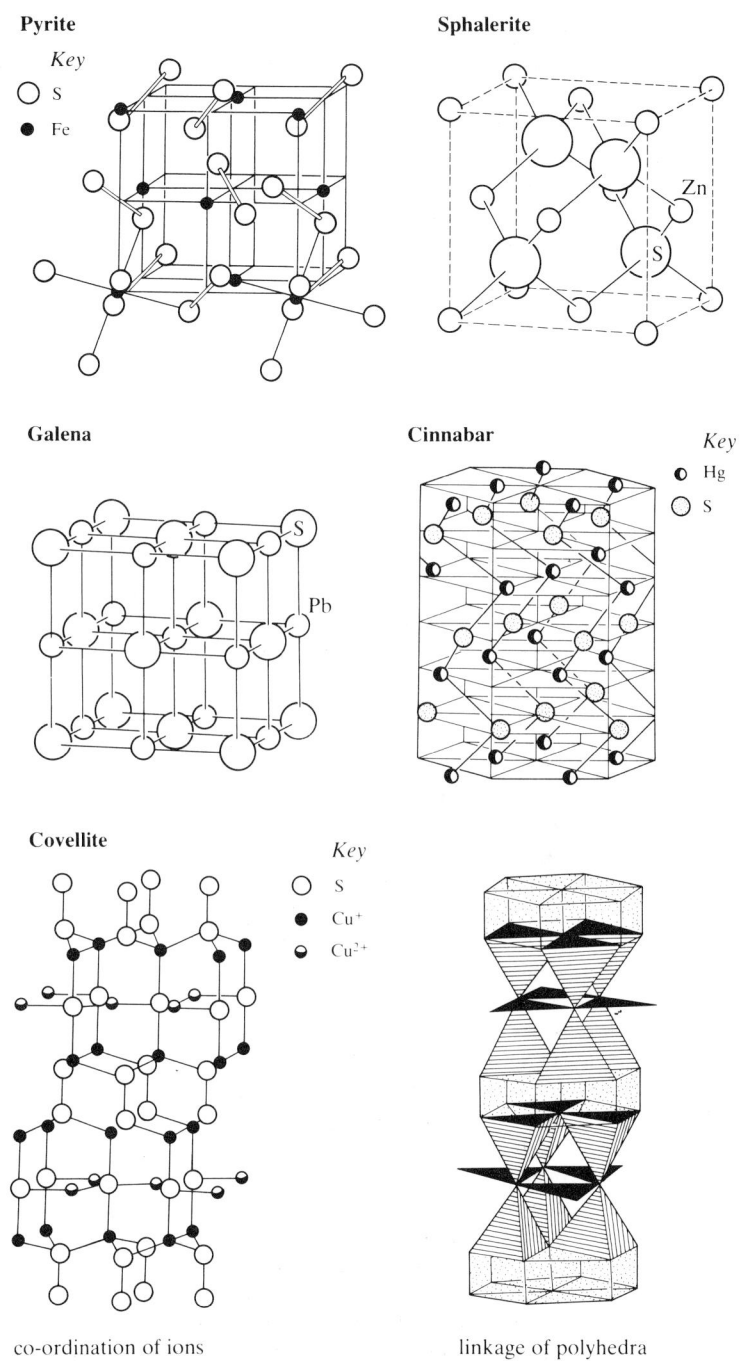

Pyrite

Key
○ S
● Fe

Sphalerite

Zn
S

Galena

S
Pb

Cinnabar

Key
◑ Hg
○ S

Covellite

Key
○ S
● Cu$^+$
◐ Cu^{2+}

co-ordination of ions

linkage of polyhedra

THE NON-SILICATES

Tetrahedrite $Cu_{10}Zn_2Sb_4S_{13}$ ½ cell
(after Pauling & Neuman 1934)

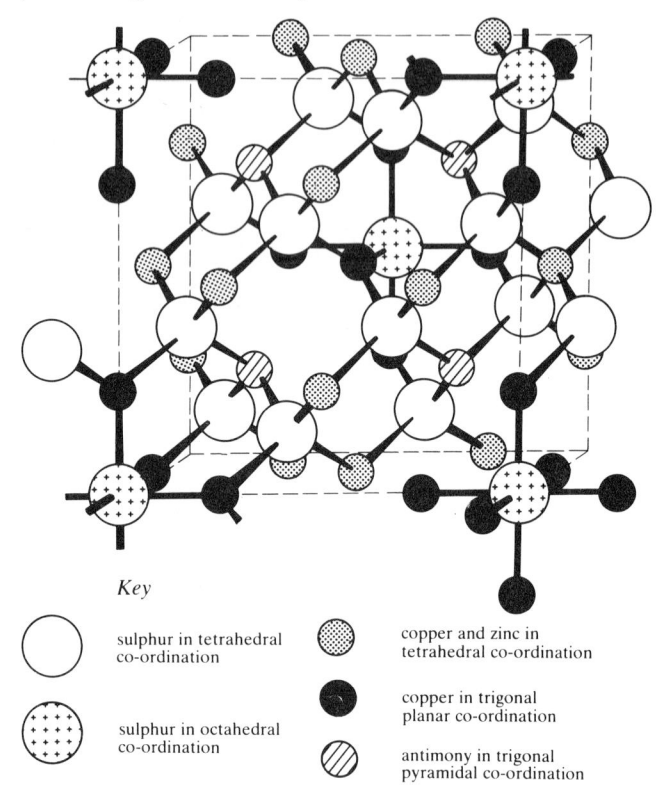

Key

○ sulphur in tetrahedral co-ordination

⊕ sulphur in octahedral co-ordination

⊙ copper and zinc in tetrahedral co-ordination

● copper in trigonal planar co-ordination

⌀ antimony in trigonal pyramidal co-ordination

and sulphur in their structures; the semi-metal is typically bonded to sulphur in trigonal pyramidal co-ordination but there is no semi-metal to metal bond as in arsenopyrite FeAsS. Two examples of sulphosalts which are relatively common are pyrargyrite Ag_3SbS_3 and tetrahedrite $(Cu,Ag)_{10}(Zn,Fe)_2(Sb,As)_4S_{13}$. The structure of tetrahedrite is illustrated in Figure 3.3 as an example of the structural complexity of sulphosalts.

Useful reviews on sulphide mineralogy are given by Vaughan and Craig (1978), Ribbe (1974) and Nickless (1968).

Arsenopyrite (mispickel)   FeAsS

Arsenopyrite is commonly non-stoichiometric and may have Fe replaced by Co. The name 'mispickel' is no longer used for arsenopyrite.

Crystals Pseudo-orthorhombic (monoclinic) with axial ratios $a:b:c =$ 1.6833 : 1 : 1.1400. Crystals are commonly prismatic [001] with twinning on: {100} and {001} giving pseudo-orthorhombic crystals; {101} giving penetration twins; or {012} giving cruciform twins.

SULPHIDES

Polished section   Cleavage $\{101\}$ is distinct. $D = 6.1$.
Arsenopyrite is white with $R \approx 52\%$, about the same as pyrite. Bireflectance is weak but anisotropy is usually quite distinct, the colours being dark blues and browns, and extinction is poor. The anisotropy is easier to observe than that of pyrite but weaker than that of marcasite.

Grain sections are often idiomorphic rhombs or lozenges or rather elongate skeletal porphyroblasts. Zonation of extinction is common and simple or hourglass twins are frequently observed. Lamellar twinning is reported. VHN = 1048–1127.

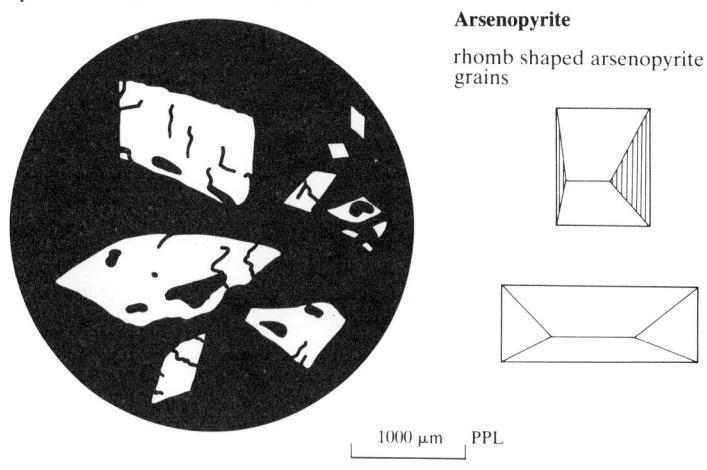

Arsenopyrite

rhomb shaped arsenopyrite grains

1000 μm   PPL

Occurrence   Arsenopyrite is considered to be typical of relatively high temperature hydrothermal veins where cassiterite, wolframite, chalcopyrite, pyrrhotite and gold are common associates. It is also found in most types of sulphide deposits.

Distinguishing features   Compared with arsenopyrite, pyrite is yellowish and cubic in morphology and marcasite is much more anisotropic.

Bornite Cu_5FeS_4

Crystals   Bornite is tetragonal (pseudo-cubic). Crystals are rare as cubes, dodecahedra or octahedra. Twinning on $\{111\}$ often results in penetration twins. $\{111\}$ is also a cleavage orientation. $D = 5.1$.

Polished section   Bornite is pinkish brown when fresh but soon tarnishes to purple or iridescent blue. With $R \approx 22\%$ it is brighter than sphalerite. Both bireflectance and anisotropy, with dark brown and grey tints, are very weak. Very fine granular aggregates appear isotropic. There is often a colour variation or zonation due to tarnishing. Multiple twinning is reported and cleavage traces in two directions are common. Chalcopyrite is commonly present as myrmekitic intergrowths or lamellae. Chalcopyrite commonly occurs along fractures. Bornite usually occurs as granular aggregates but is often intergrown with other Cu + Fe + S minerals. VHN = 97–105.

THE NON-SILICATES

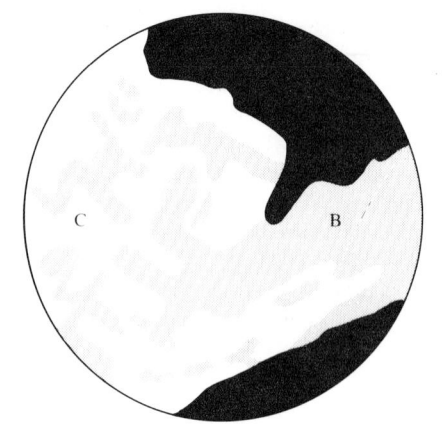

Bornite

myrmekitic intergrowth of bornite (B) and chalcopyrite (C)

300 μm PPL

Occurrence Bornite is usually associated with other Cu + Fe + S minerals in the 'secondary environment'.

It can result from unmixing of high temperature Cu + Fe + S solid solutions on cooling.

Distinguishing features Compared with bornite, pyrrhotite is lighter brown and distinctly anisotropic; they rarely occur together.

Chalcocite Cu_2S
Digenite Cu_9S_5

Ramdohr (1969) states that 'what has hitherto been considered as "chalcocite" with the formula Cu_2S is a great number of semi-independent minerals and solid solutions, whose relationships are not yet fully understood and for which there are diverse interpretations'. Care must therefore be taken when examining samples reportedly containing chalcocite.

Crystals Chalcocite Cu_2S is orthorhombic, $a:b:c = 0.5822:1:0.9701$. Digenite Cu_9S_5 is cubic. Both minerals are usually massive. $D = 5.77$.

Polished section Chalcocite appears bluish light grey with $R \approx 32\%$, whereas digenite is light grey to bluish light grey with $R \approx 22\%$. Chalcocite is weakly anisotropic with pinkish, greenish grey or brownish tints. Digenite is isotropic.

Both minerals occur in granular aggregates and are commonly in intergrowths with each other or other Cu + Fe + S minerals. Lancet shaped inversion twinning indicates cooling from the high temperature hexagonal polymorph through 103 °C to the orthorhombic polymorph. Cleavage traces may be observed and are enhanced on weathering. VHN: 68–98 chalcocite, 56–67 digenite.

Occurrence Digenite is indicative of higher temperatures and higher sulphur activity than chalcocite. Both minerals are associated with other copper and iron sulphides, especially covellite, in low temperature hydrothermal veins and in the 'secondary environment'. They occur in cupriferous, red-bed sedimentary rocks and are widespread as replacement minerals.

SULPHIDES

Distinguishing features — Compared with chalcocite, djurleite $Cu_{1.96}S$ (orthorhombic) is very similar; sphalerite is slightly darker, isotropic, and often shows internal reflections; and tetrahedrites are less blue, harder and isotropic.

Notes — Copper sulphide minerals are complex owing to the variation in crystallographic and optical properties with slight changes in Cu:S ratio. Their colour changes readily owing to surface damage during polishing as well as to tarnishing.

Chalcopyrite $CuFeS_2$

Incorporation of many other elements (e.g. Ni, Zn, Sn) is possible at high temperatures in the cubic polymorph, which has a range in composition in the Cu + Fe + S system. Unmixing occurs on cooling, resulting in inclusions in chalcopyrite.

Crystals — Chalcopyrite is tetragonal, $a:c = 1:1.9705$. Crystals are commonly scalenohedral or tetrahedral in appearance (Fig. 3.4). Twinning is common on $\{112\}$ and $\{012\}$ and cleavage is $\{011\}$. $D = 4.28$.

Thin section — Chalcopyrite is opaque but alteration leads to associated blue green staining or associated secondary hydrous copper carbonates which are blue to green in colour.

Polished section — Chalcopyrite is yellow and tarnishes to brownish yellow. R = 42–46%, slightly less than pyrite and similar to galena. Anisotropy is weak with dark brown and greenish grey tints, and is often not visible.

Chalcopyrite usually occurs as irregular or rounded grains. It is common as rounded inclusions or in fractures in other sulphides, especially pyrite and sphalerite. Colloform masses of chalcopyrite have been reported. Simple and multiple twinning is common and cleavage traces are sometimes observed. Several phases may be present in chalcopyrite as exsolved blebs, lamellae or stars (e.g. ZnS) and indicate a high temperature origin. VHN = 186–219.

Occurrence — Chalcopyrite is a common accessory mineral in most types of ore deposit as well as in igneous and metamorphic rocks. It is the major primary copper mineral in prophyry copper deposits and it occurs, with bornite, in the stratiform sulphide deposits of the Copperbelt. Chalcopyrite appears to be a relatively mobile mineral in ore deposits and commonly replaces and veins other minerals, especially pyrite.

Distinguishing features — Compared with chalcopyrite, pyrite is white, much harder and commonly idiomorphic, and gold is much brighter but may be yellower or whiter. Small isolated grains of pyrite and chalcopyrite can be very similar in appearance.

Figure 3.4 Typical chalcopyrite crystal.

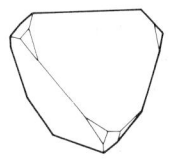

Cinnabar HgS

Crystals Cinnabar is trigonal, $a:c = 1:2.2905$, and occurs as thick tabular $\{0001\}$ or prismatic $[10\bar{1}1]$ crystals. There is a $\{0001\}$ twin plane and perfect $\{10\bar{1}1\}$ cleavage. $D = 8.09$.

Thin section Cinnabar is deep red. Refractive index values ($\lambda = 598$ nm) are $n_o = 2.905$ and $n_e = 3.256$.

Polished section Cinnabar is light grey to bluish light grey, weakly pleochroic, with $R_o = 28\%$ and $R_e = 29\%$. Anisotropy is moderate with greenish grey tints, but these are often masked by abundant deep red internal reflections.

Cinnabar occurs as granular aggregates and idiomorphic crystals. Deformation multiple twinning may be present. As a result of variation in polishing hardness with orientation, granular aggregates may resemble a two phase intergrowth at first glance.

Occurrence Cinnabar is rare, occurring in low temperature hydrothermal veins, impregnations and replacement deposits often associated with recent volcanics. It often replaces quartz and sulphides and is associated with native mercury, mercurian tetrahedrite–tennantite, stibnite, pyrite and marcasite in siliceous gangue. VHN = 51–98.

Distinguishing features Compared with cinnabar, hematite is brighter, harder and has very rare internal reflections; pyrargyrite is very similar but with less intense internal reflections; and cuprite Cu_2O is bluish grey, harder and usually associated with native copper.

Notes Metacinnabarite is a high temperature cubic polymorph of HgS. It occurs as grains within cinnabar and is slightly darker, isotropic, lacks internal reflections and is softer than cinnabar.

Cobaltite CoAsS

Cobaltite may contain significant amounts of Fe and Ni in solid solution.

Crystals Cobaltite is orthorhombic (pseudo-cubic). It commonly occurs in cubes or pyritohedrons but may be octahedral. There is a perfect $\{001\}$ cleavage. $D = 6.0$–6.3.

Polished section Cobaltite is pinkish white with $R \approx 53\%$, similar to pyrite. Both bireflectance and anisotropy, with brownish to bluish tints, are weak. Cobaltite is often idiomorphic and of 'cubic' morphology. It may be granular or skeletal. Colour zonation has been observed and complex fine lamellar twinning and cleavage traces may be present. VHN = 1176–1226.

Occurrence It is associated with Cu + Fe + S and Co + Ni + As minerals in high to medium temperature deposits in veins and as disseminations.

Distinguishing features Compared with cobaltite, pyrite is yellowish and harder.

Covellite CuS

Covelline is an alternative name recommended by the International Mineralogical Association.

SULPHIDES

Crystals   Covellite is hexagonal, $a:c = 1:1.43026$. It occurs as platy $\{0001\}$ crystals with a perfect $\{0001\}$ basal cleavage. $D = 4.6$.

Thin section   Covellite is greenish in very thin flakes.

Polished section   Covellite is blue and strongly pleochroic from blue to bluish light grey, except in basal sections which remain blue. $R_o = 7$ and $R_e = 22\%$. Anisotropy is very strong with bright 'fiery' orange colours.

Covellite occurs as idiomorphic platy crystals and flakes as well as rather 'micaceous' aggregates. The good basal cleavage, parallel to the length of grains, is often deformed. VHN = 69–78.

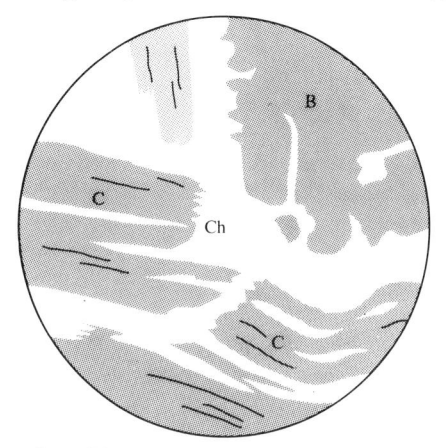

Covellite

basal section (B) and cross sections of covellite showing cleavage traces (C): chalcocite (Ch) is replacing the covellite

500 μm   PPL

Occurrence   Covellite commonly occurs as a 'secondary' mineral after Cu + Fe + S minerals, often in the zone of secondary enrichment.

Distinguishing features   Covellite is easy to identify. Digenite is blue but neither pleochroic nor anisotropic.

Notes   Blaubleibender covellite $Cu_{1+x}S$ occurs with covellite and is identical in appearance except under oil immersion, when:

covellite R_o = reddish purple R_e = bluish grey
blaubleibender covellite R_o = deep blue R_e = bluish grey

Galena   PbS

Galena may possibly contain some Se, Te, Ag, Sb, Bi, As in solid solution but usually only in trace amounts.

Crystals   The crystallographic symmetry of galena is cubic and crystals are commonly cubic, cubo-octahedral and (less often) octahedral in shape (Fig. 3.5). Twinning on $\{111\}$ is common and lamellar twinning may occur on $\{114\}$. There is a perfect $\{001\}$ cleavage. $D = 7.58$.

Figure 3.5
Typical galena crystals.

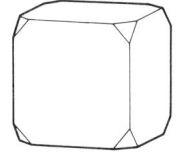

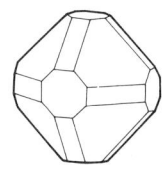

THE NON-SILICATES

Polished section Galena is white, sometimes with a very slight bluish tint. $R = 43\%$, which makes it darker than pyrite. It is isotropic but sometimes very weakly anisotropic.

Galena commonly has cubic morphology in vein and replacement deposits. It is often interstitial to other sulphides and occurs in microfractures. Internal grain boundaries of granular aggregates are enhanced by excessive polishing. Triangular cleavage pits are characteristic of galena and it is often altered along cleavage traces. Many minerals occur as inclusions, but especially sulphosalts of Pb, Ag with Sb or As. VHN = 71–84.

Galena

cleavage pits including the characteristic triangular cleavage pits in galena

500 μm   PPL

Galena

myrmekitic intergrowth of galena (white) and tennantite (light grey): limestone and dolomite rhombs are dark grey

300 μm   PPL

Occurrence Galena is common in hydrothermal vein and replacement deposits in many rock types but especially limestones. It is also common in some young (Proterozoic and Phanerozoic) stratiform massive sulphide deposits. Sphalerite is a common associate.

Distinguishing features Compared with galena, some Pb + Sb + S minerals are similar but these are usually distinctly anisotropic.

Marcasite FeS$_2$

Crystals Marcasite is orthorhombic, $a:b:c = 0.8194:1:0.6245$. It is commonly tabular $\{010\}$ but may be pyramidal. Aggregates are often globular or stalactitic. Twinning on $\{101\}$ is common, often repeated, producing cockscomb pseudo-hexagonal shapes. Cleavage on $\{101\}$ is distinct. $D = 4.88$.

Thin section Marcasite is opaque, but because of ready oxidation a brown staining of limonite is often associated with it.

Polished section Marcasite is white, slightly yellowish. There is a weak pleochroism, ∥a pinkish white and ∥b and ∥c yellowish white. $R = 49-55\%$, very close to pyrite. The strong anisotropy of marcasite in very bright bluish and greenish greys and browns is one of the most distinctive features.

Occurrence Marcasite often appears as lath-shaped crystals in radiating aggregates of twins. Colloform aggregates with pyrite are common. Lamellar twinning and cleavage pits may be present. VHN = 941–1288.

Distinguishing features Marcasite often occurs as concretions in sedimentary rocks. It is usually associated with pyrite in low temperature sulphide deposits.

Compared with marcasite, pyrite is yellower, slightly softer and weakly anisotropic or isotropic; pyrrhotite is darker, brownish, softer and has a weaker anisotropy; and arsenopyrite is whiter, brighter and has a weaker anisotropy.

Molybdenite MoS$_2$

Molybdenite may contain Rh.

Crystals Molybdenite is hexagonal, $a:c = 1:3.815$. Having a layer structure, it commonly has a hexagonal tabular or a short barrel shaped prismatic habit. It is commonly foliated massive or in scales. There is a perfect basal $\{0001\}$ cleavage. $D = 4.7$.

Thin section Molybdenite is opaque in the visible but it is transparent and uniaxial $-$ve in the infrared.

Polished section Molybdenite is bireflecting with $R_o = 37\%$ (white, less bright than galena) and $R_e = 15\%$ (grey, similar to sphalerite). Anisotropy is very strong with slightly pinkish white tints. Extinction is parallel to cleavage (the brighter R_o orientation) but is often undulatory because of deformation.

It occurs as flakes or platelets with hexagonal basal sections. Well-developed basal cleavage often results in a poor polish, especially on grains which have their cleavage parallel to the polished surface. Deformation twinning-like structure is related to buckling of cleavage. VHN = 16–19 ⊥ cleavage, 21–28 ∥ cleavage.

Occurrence Molybdenite is found in high temperature hydrothermal veins and quartz pegmatites, with Bi, Te, Au, Sn and W minerals. It also occurs in porphyry copper style deposits. It is an accessory mineral in acid igneous rocks and occasionally is a detrital mineral.

THE NON-SILICATES

Distinguishing features Compared with molybdenite, tungstenite WS$_2$ is very similar; graphite is morphologically similar but much darker; and tetradymite Bi$_2$Te$_2$S is brighter.

Notes Molybdenite polishes poorly because of smearing.

Pentlandite (Fe,Ni)$_9$S$_8$

Pentlandite usually contains about equal amounts of Fe and Ni. It often contains Co and sometimes Cu or Ag in solid solution.

Crystals Pentlandite is cubic but rarely occurs as well shaped crystals. There is no cleavage but a parting on $\{111\}$. $D = 5.0$.

Polished section Pentlandite is very slightly yellowish white (cream) with $R = 52\%$ which is similar to pyrite. It is isotropic.

It occurs commonly as 'flame' lamellae in pyrrhotite and as veinlets or xenomorphic grains associated with pyrrhotite. The octahedral parting $\{111\}$ is often well developed, resulting in triangular cleavage pits. Alteration also takes place along this parting. VHN = 202–230.

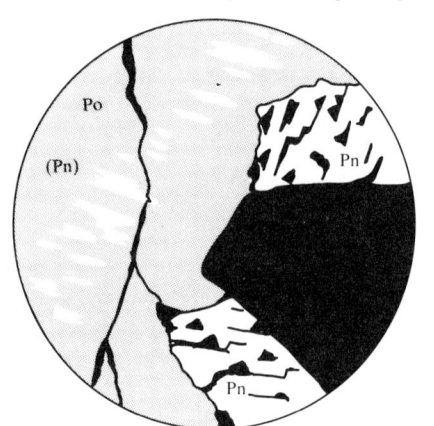

Pentlandite

exsolved pentlandite (Pn) 'flames'in pyrrhotite (Po): also coarse pentlandite showing pits due to octahedral parting

200 μm   PPL

Occurrence Pentlandite, usually associated with pyrrhotite and other Cu + Ni + Fe + S phases, is common in mafic igneous rocks, e.g. norites, and some massive sulphide deposits.

Distinguishing features Compared with pentlandite, pyrite is yellowish, often weakly anisotropic and harder, and pyrrhotite is darker, brownish, anisotropic and slightly harder.

Pyrargyrite Ag$_3$SbS$_3$

Pyrargyrite and proustite Ag$_3$AsS$_3$ are known as the 'ruby silvers' because they are translucent with a deep red colour. Extensive solid solution occurs between the two minerals.

Crystals Pyrargyrite is triagonal, $a:c = 1:0.7892$ and proustite is trigonal, $a:c = 1:0.8039$. Both minerals are commonly prismatic $[0001]$ with twinning, sometimes complex, on $\{10\bar{1}4\}$. There is a distinct $\{10\bar{1}1\}$ cleavage. $D = 5.85$ ($D = 5.57$ proustite).

SULPHIDES

Thin section   Both minerals are deep red, uniaxial −ve.

Polished section   Both minerals are light grey, often slightly bluish. $R = 28–31\%$ (pyrargyrite) and $R = 25–28\%$ (proustite), which makes them similar in brightness to tetrahedrite. Bireflectance is distinct and anisotropy strong in greys. Red internal reflections are common and more abundant in proustite.

Both minerals occur as isolated crystals but are common as inclusions in galena. Simple and multiple twinning may be present. VHN: pyragyrite 50–97 ⊥ cleavage, 98–126 ∥ cleavage; proustite VHN 109–135.

Occurrence   Pyrargyrite is more common than proustite. They are associated with other sulphosalts, especially tetrahedrite–tennantite, in low temperature Pb + Zn mineralisation and Ag + Ni + Co veins. The ruby silvers and similar Ag minerals may be significant silver carriers in base metal mineralisation.

Distinguishing features   There are some rare complex sulphides which resemble the ruby silvers. Cinnabar is quite similar but the anisotropy tints are greenish grey.

Pyrite FeS_2

Pyrite may contain some Ni or Co. Auriferous pyrite probably contains inclusions of native gold and cupriferous pyrite probably contains inclusions of chalcoypyrite.

Crystals   Pyrite is cubic, crystals most commonly being modifications of cubes (Fig. 3.6). The $\{011\}$ twin plane and $[001]$ twin axis produce penetration twins. There is a poor $\{001\}$ cleavage. $D = 5.01$.

Thin section   Pyrite is opaque, often occurring as euhedral crystals or aggregates of small rounded grains. Alteration to limonite results in brownish or reddish coloured rims or brown staining.

Polished section   Pyrite is white, often with a slight yellowish tint especially in small grains. $R = 54\%$, resulting in pyrite usually appearing very bright. It is only ideally isotropic in (111) sections, and the weak anisotropy in very dark green and brown can usually be seen in well polished grains.

Pyrite is usually idiomorphic but is occasionally intergrown with other sulphides, e.g. sphalerite. Grains are often cataclased. Framboidal pyrite is common in sedimentary rocks. Growth zoning in pyrite is enhanced by etching. Zonation of inclusions is common. Inclusions of other sulphides are common, e.g. chalcopyrite, pyrrhotite. Fractures in pyrite often contain introduced sulphides, e.g. chalcopyrite, galena. $VHN = 1027–1240$.

Figure 3.6 Typical pyrite crystals.

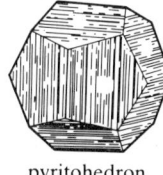

pyritohedron

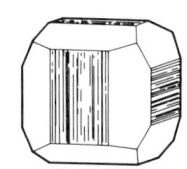

THE NON-SILICATES

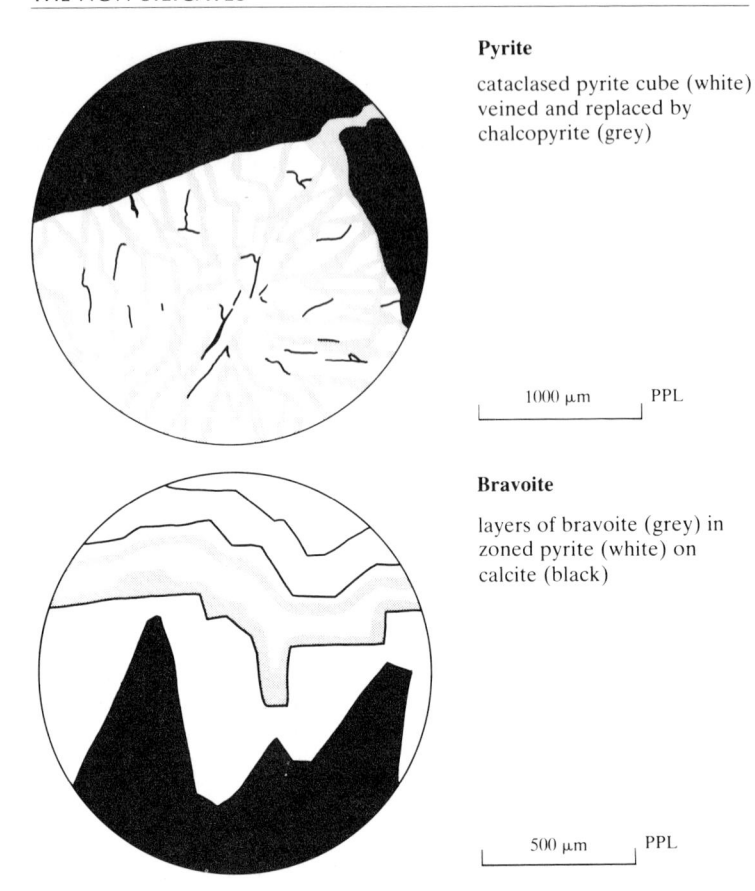

Pyrite

cataclased pyrite cube (white) veined and replaced by chalcopyrite (grey)

1000 μm PPL

Bravoite

layers of bravoite (grey) in zoned pyrite (white) on calcite (black)

500 μm PPL

Occurrence Pyrite is a common sulphide occurring in most rocks and ores. Organic material, carbonates and quartz are all readily replaced by pyrite.

Distinguishing features Compared with pyrite, marcasite is whiter and strongly anisotropic; chalcopyrite is distinctly yellow and much softer; arsenopyrite is whiter and tends to form rhomb shapes; and pentlandite is whiter, softer and often shows alteration along octahedral partings.

Notes Melnikovite is poorly crystallised colloform iron sulphide which appears brownish and porous. It probably consists of FeS_2 and hydrous FeS. It tarnishes rapidly.

Bravoite is nickeliferous pyrite $(Fe,Ni)S_2$, often with some Co. It is similar to pyrite but brownish, slightly darker and anisotropic. It usually occurs as idiomorphic centres or as layers in zoned pyrite.

Pyrrhotite $Fe_{1-x}S$

Pyrrhotite may contain some Ni, Co or Mn. It is cation deficient relative to the stoichiometric mineral troilite FeS. Nickeliferous pyrrhotite probably contains pentlandite.

SULPHIDES

Pyrrhotine is an alternative name recommended by the International Mineralogical Association.

Crystals Both monoclinic and hexagonal, $a:c = 1:1.6502$, varieties of pyrrhotite occur, and these are commonly intergrown. Crystals are commonly tabular to platy with twinning on $\{10\bar{1}2\}$. There is no cleavage. $D = 4.6$.

Polished section Pyrrhotite is brownish or pinkish white with a weak but usually visible pleochroism. $R \approx 40\%$ with R_o being darker and R_e being lighter in the case of hexagonal pyrrhotite. Anisotropy is strong with yellowish, greenish or bluish grey tints.

Pyrrhotite is usually xenomorphic, often occurring as polycrystalline aggregates or as inclusions in pyrite. Multiple twinning, often spindle-shaped due to deformation, is common. Exsolved lamellae (or flames) of white pentlandite are common. VHN = 230–318.

Occurrence The presence of pyrrhotite indicates a relatively low S availability. It is common in igneous rocks, metamorphic rocks and stratiform massive Cu + Fe + S deposits. It forms in the reaction

$$FeS_2 \rightarrow Fe_{1-x}S + S\uparrow$$

in contact metamorphic aureoles. In veins, it is usually taken to indicate precipitation from relatively high temperature, acid, reducing solutions.

Distinguishing features Hexagonal and monoclinic pyrrhotite are not easily distinguished in polished section. A magnetic colloid may be used to stain monoclinic pyrrhotite (Craig and Vaughan 1981). Compared with pyrrhotite, ilmenite is darker and harder; bornite is browner (soon tarnishing to purple) and essentially isotropic; and chalcopyrrhotite (rare) is isotropic and browner than pyrrhotite.

Notes Pyrrhotite alters readily along irregular fractures to a mixture of iron minerals including marcasite, pyrite, magnetite and limonite. Although rare in sedimentary rocks and common in metamorphosed equivalents, especially near synsedimentary stratiform sulphide deposits, pyrrhotite is not thought to be necessarily a metamorphic mineral formed by breakdown of pyrite. It may be of hydrothermal exhalative origin and could persist in sea-floor sediments provided the sulphur availability was low.

Sphalerite ZnS

Sphalerite usually contains Fe and sometimes Cd, Mn or Hg in solid solution.

Crystals Sphalerite is cubic. It has the diamond structure (see Fig. 3.3) but is more complex than one might suspect; there are many polytypes. Crystals are commonly tetrahedral and dodecahedral (Fig. 3.7). Twinning about the $[111]$ axis leads to simple and complex twins. There is a perfect $\{011\}$ cleavage. $D = 3.9$.

Thin section Pure ZnS is transparent and colourless but sphalerite is opaque when iron rich. It has very high relief and is usually yellow to brownish in

THE NON-SILICATES

Figure 3.7 Typical sphalerite crystals.

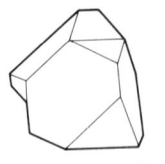

spinel-type twin

colour with dark brown bands due to Fe zonation. Oxidation of iron-bearing varieties leads to brown staining, especially in fractures. Sphalerite is isotropic but is sometimes anomalously anisotropic, revealing fine lamellar twinning probably due to stacking polytypes. At $\lambda = 589$ nm, $n = 2.369$ (pure ZnS), 2.40 (5.46% Fe), 2.43 (10.8% Fe) and 2.47 (17.06% Fe).

Polished section Sphalerite is grey with $R = 17\%$. It is darker than most ore minerals but brighter than the gangue minerals. It is isotropic. Pure ZnS has abundant internal reflections but, with increasing Fe content, opacity increases and internal reflections become fewer and brownish or reddish.

Sphalerite is rarely idiomorphic. It usually occurs as rounded grains in aggregates. It also is found as zoned colloform masses. Irregular fractures are common and the cleavage often results in severe pitting. Multiple twinning is often visible. Zonation of iron, seen as brown bands in transmitted light or by internal reflection, does not visibly change brightness. Sphalerite usually contains inclusions, especially of chalcopyrite, as blebs or lamellae. VHN = 186–209.

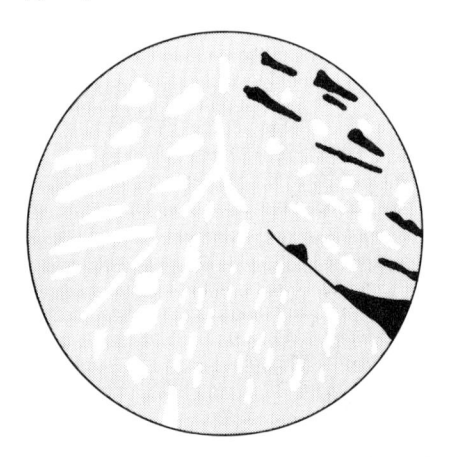

Sphalerite

chalcopyrite (white) blebs and interstitial veinlets in sphalerite (grey)

200 μm   PPL

Occurrence Sphalerite is common in stratabound, vein and massive sulphide deposits. Sphalerite, typically very low in Fe content, also occurs with galena, pyrite and chalcopyrite in calcareous nodules or veinlets probably of diagenetic origin. Fe-rich sphalerite often occurs with pyrrhotite, as it is the activity of FeS rather than the abundance of Fe that controls the iron content of sphalerite. Sphalerite is often associated with galena.

SULPHIDES

Distinguishing features Compared with sphalerite, magnetite is often pinkish, harder and never has internal reflections; limonite is bluish grey, usually has reddish internal reflections and is usually replacing iron-bearing minerals; and tetrahedrite is brighter, greenish or bluish grey and only very rarely shows internal reflections.

Notes Wurtzite (hexagonal ZnS) is very similar to sphalerite in polished section, but it is rare.

Stibnite Sb_2S_3

Crystals Stibnite is orthorhombic with $a:b:c = 0.9926:1:0.3393$. Crystals are usually prismatic $[001]$, often slender to acicular. Twinning on $\{130\}$ is rare. There is a perfect $\{010\}$ cleavage and imperfect $\{100\}$ and $\{110\}$ cleavages. $D = 4.63$.

Thin section Stibnite is opaque. However, it is transparent using infra-red transmitted light.

Polished section Stibnite has a pronounced bireflectance with $R = 30$ to 40%. It is light grey ||a, brownish light grey ||b and white ||c. The anisotropy is very strong, with tints ranging from light bluish grey to brown. Extinction is straight.

Stibnite often occurs as acicular or bladed crystals or granular aggregates. The usually well developed cleavage traces are deformed. Deformation twinning is common. VHN = 42–109.

Stibnite

bladed stibnite grains showing distinct bireflectance (white to grey) and cleavage traces: in quartz (black)

500 μm   PPL

Occurrence Stibnite is found in low temperature hydrothermal veins, usually with quartz. It is associated with complex Sb-bearing and As-bearing sulphides, pyrite, gold and mercury.

Distinguishing features Compared with stibnite, hematite has a smaller bireflectance, weaker anisotropy, is harder and lacks cleavage. Some lead-antimony sulphides are very similar to stibnite.

THE NON-SILICATES

Tetrahedrite $Cu_{10}(Zn,Fe)_2Sb_4S_{13}$
Tetrahedrite exhibits extensive chemical substitution and often contains Ag, Hg and As but only rarely Cd, Bi and Pb. The arsenic end member is tennantite $Cu_{10}(Zn,Fe)_2As_4S_{13}$. Silver-rich tetrahedrite is known as freibergite. Tetrahedrite–tennantites were often formerly called fahlerz.

Crystals Tetrahedrite is cubic and occurs as modified tetrahedra. Twinning on the axis [111] is often repeated. There are also penetration twins. There is no cleavage. $D \approx 5.0$.

Thin section Tetrahedrites are usually opaque, but iron-free and arsenic-rich varieties transmit some red light.

Polished section Tetrahedrite is light grey, sometimes appearing slightly greenish, bluish or brownish. With $R \approx 31\%$ it is darker than galena but brighter than sphalerite. It is usually isotropic but may be weakly anisotropic. Very scarce red internal reflections have been reported from tennantite.

Tetrahedrite is rarely idiomorphic. It is usually in the form of rounded grains or polycrystalline aggregates. It forms myrmekitic intergrowths with other sulphides, e.g. galena, chalcopyrite. Zonation of Sb/As and Fe/Zn is commonly detected on microanalysis but is not visible in polished section. Irregular fracturing is common. Inclusions, especially of chalcopyrite, are common. VHN = 320–367.

Occurrence Tetrahedrite commonly occurs associated with galena in lead + zinc deposits, although it is inexplicably abundant in some and absent in others. Tennantite is common in porphyry copper mineralisation.

Distinguishing features Compared with tetrahedrites, sphalerite is darker, harder, has a good cleavage and usually shows internal reflections. Many complex sulphides (sulphosalts) are similar at first glance to tetrahedrite, but most of these are anisotropic.

Notes The various chemical varieties of tetrahedrite cannot be identified with any certainty in polished section without resorting to microanalysis.

3.4 Oxides

Oxides are minerals that contain one or more metals and oxygen; quartz SiO_2 is usually excluded from the group. The reader is referred to Rumble (1976) for a review of the oxides. As most silicates in igneous and metamorphic rocks consist essentially of silica plus metal oxides, free oxides can be considered as forming if metal oxides are present surplus to the needs of silicates. Alternatively, they may form if rocks are silica deficient, as is usually the case when periclase MgO and corundum Al_2O_3 are found, or they may form if the metal is 'inappropriate' for a silicate structure, e.g. cassiterite SnO_2. The two following groups, the iron–titanium oxides and the spinels, overlap to a certain degree but will be outlined briefly because they contain the most common oxides. Note that all the oxides are listed alphabetically.

OXIDES

Iron-titanium oxides

Minerals with chemical compositions of essentially iron, titanium and oxygen are of widespread occurrence in rocks of all types. Their identification is important because much can be learnt about the crystallisation history of the host rock. Rumble (1976) states: 'The oxide minerals are of great value in deducing the conditions of metamorphism; indeed, their value is out of all proportion to their modal abundance in typical rocks, for they simultaneously record information on both the ambient temperature and the chemical potential of oxygen during metamorphism.' The same statement may be applied to igneous rocks.

The triangular diagram Figure 3.8 shows the Fe–Ti–O minerals. Although magnetite, ilmenite and hematite are usually considered to be the common examples, precise identification may be difficult due to extensive chemical substitution within the Fe–Ti–O system as well as the presence of Cr, Mn, Mg and Al in these minerals. The Fe–Ti–O minerals often occur in intergrowths, frequently submicroscopic, which result from cooling and oxidation/reduction.

In typical basaltic igneous rocks there are two primary oxide minerals, ferrianilmenite and titanomagnetite. On slow cooling ferrianilmenite may become ilmenite with hematite lamellae, whereas titanomagnetite may produce lamellae of ilmenite before breaking down to a fine intergrowth of ulvospinel and magnetite. Regionally metamorphosed sedi-

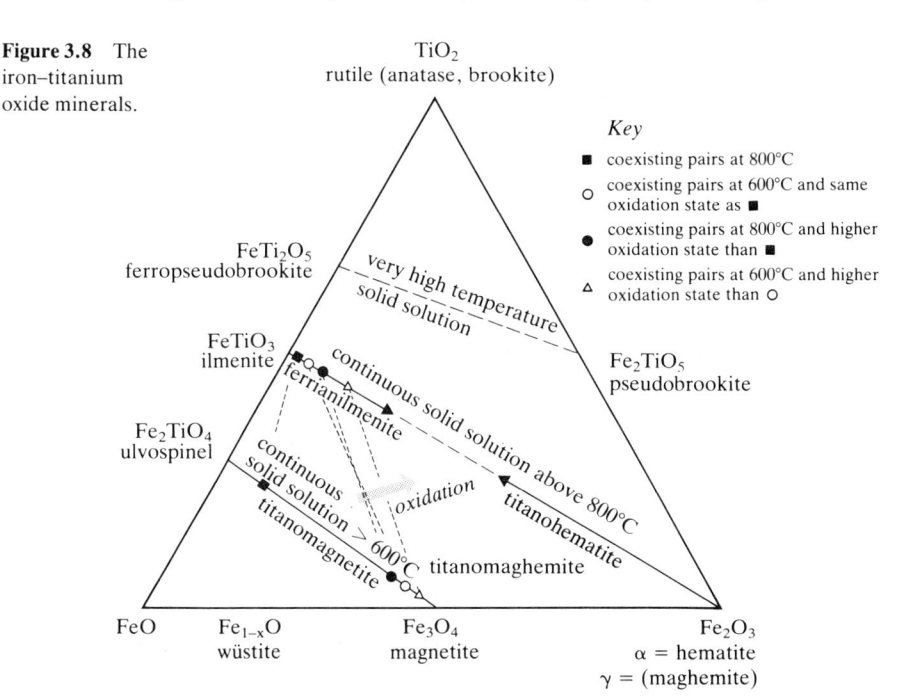

Figure 3.8 The iron–titanium oxide minerals.

Key
- ■ coexisting pairs at 800°C
- ○ coexisting pairs at 600°C and same oxidation state as ■
- ● coexisting pairs at 800°C and higher oxidation state than ■
- △ coexisting pairs at 600°C and higher oxidation state than ○

Triangle vertices:
- TiO$_2$ rutile (anatase, brookite)
- FeO / Fe$_{1-x}$O wüstite / Fe$_3$O$_4$ magnetite / Fe$_2$O$_3$ α = hematite, γ = (maghemite)

Labels along sides:
- FeTi$_2$O$_5$ ferropseudobrookite
- FeTiO$_3$ ilmenite
- Fe$_2$TiO$_4$ ulvospinel
- Fe$_2$TiO$_5$ pseudobrookite
- very high temperature solid solution
- continuous solid solution ferrianilmenite
- continuous solid solution above 800°C titanohematite
- continuous solid solution titanomagnetite
- oxidation > 600°C titanomaghemite

ments typically contain assemblages of almost pure magnetite ± ferrianilmenite (possibly with exsolved hematite) ± titanohematite (possibly with exsolved ilmenite) ± rutile. Greenschist facies rocks may contain a magnetite + rutile assemblage which gives way to a titanohematite + ferrianilmenite assemblage in amphibolite facies.

The following are noted for Figure 3.8:

(a) The TiO_2 polymorphs are:

rutile, tetragonal $c/a < 1$
anatase, tetragonal $c/a > 1$ (metastable?) ⎱ found in low
brookite, orthorhombic (metastable?) ⎰ temperature hydrothermal environment

(b) Ferropseudobrookite is only stable above 1100 °C and is very rare.
(c) Pseudobrookite is only stable above 585 °C and is rare (in high temperature contact metamorphosed rocks).
(d) The ilmenite–hematite solid solution series is complete above about 800 °C.
(e) The ulvospinel–magnetite solid solution series is complete above about 600 °C.
(f) Magnetite and rutile can coexist only below about 400 °C.
(g) From 1100 °C to 600°C Ti-rich ferrianilmenite coexists with Ti-poor titanomagnetite; the exact composition of the coexisting pair depends on oxygen fugacity as well as temperature (the Buddington and Lindsley (1964) magnetite–ilmenite geothermometer oxygen barometer). The dotted lines show how the compositions (approximate) of coexisting pairs depend on temperature and oxygen fugacity.
(h) Oxidation of titanomagnetite at relatively high temperatures results in exsolution lamellae of ferrianilmenite in the (111) orientation in magnetite, and this oxidation can result from cooling alone. Similarly, reduction of ferrianilmenite results in titanomagnetite lamellae in the (0001) orientation of ilmenite.
(i) Titanomaghemites form at low temperatures ($< 600 °C$) by non-equilibrium oxidation of titanomagnetites; they are cation deficient and have a wide range in composition.
(j) Hemo-ilmenite is a ferrianilmenite host with titanohematite lamellae.
(k) Ilmeno-hematite is a titanohematite host with ferrianilmenite lamellae.
(l) The *bulk* composition of coexisting hemo-ilmenite and ilmeno-hematite grains depends on temperature.
(m) Wüstite is cation deficient relative to FeO. It is very rare as it is stable only above 570 °C at low oxygen fugacities.

OXIDES

The spinel group

The general unit cell formula of the spinels is $R_8^{2+}R_{16}^{3+}O_{32}$, where R^{2+} and R^{3+} stand for divalent and trivalent cations respectively but the formula is usually simplified to R_3O_4. All spinels are cubic but there are two structural types with differing distributions of the cations:

Normal spinels
 R_8^{2+} in fourfold tetrahedral co-ordination with oxygen
 R_{16}^{3+} in sixfold octahedral co-ordination with oxygen

Inverse spinels
 R_8^{3+} in fourfold tetrahedral co-ordination with oxygen
 R_8^{2+} and R_8^{3+} in sixfold octahedral co-ordination with oxygen

Most natural spinels have an intermediate structure (see Fig. 3.9). In the spinel structure, oxygen is O^{2-} in tetrahedral co-ordination.

The spinels are normal valence compounds in that the total cation charge balances the total anion charge. Divalent R^{2+} cations include Mg^{2+}, Fe^{2+}, Zn^{2+}, Mn^{2+} and Ni^{2+}, and R^{3+} cations include Al^{3+}, Fe^{3+} and Cr^{3+}. One way of representing the extensive solid solution in spinels is shown in Figure 3.10.

It is possible for Ti^{4+} (and V^{4+}) to enter the structure due to a coupled substitution of the type $2Fe^{3+} \rightleftharpoons Fe^{2+} + Ti^{4+}$. The inverse spinel structure of maghemite γ-Fe_2O_3 supports a cation site vacancy which is produced by the substitution $3Fe^{2+} \rightleftharpoons 2Fe^{3+} + [\ ]$; the formula may be written

$$Fe_3^{3+}(tetr.)Fe_3^{3+}(oct.)Fe_2^{3+}[\ ](oct.)O_{12}$$

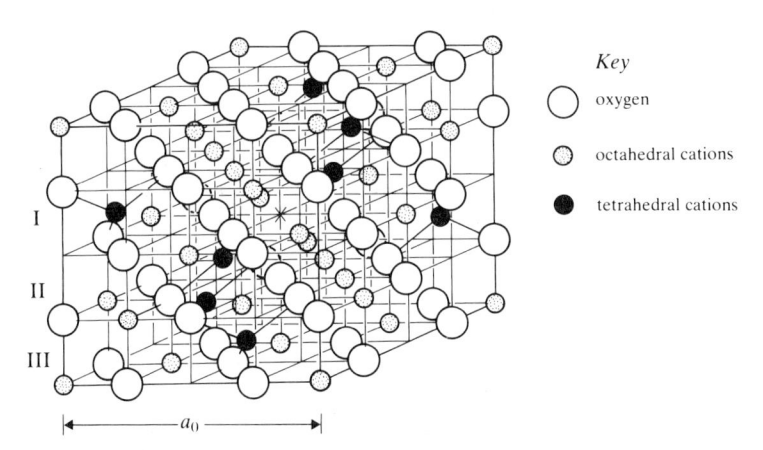

Figure 3.9 The spinel unit cell, orientated so as to emphasise the (111) planes. Atoms are not drawn to scale; the circles simply represent the centres of atoms (after Lindsley, in Rumble 1976).

THE NON-SILICATES

Figure 3.10 Solid solution in spinels.

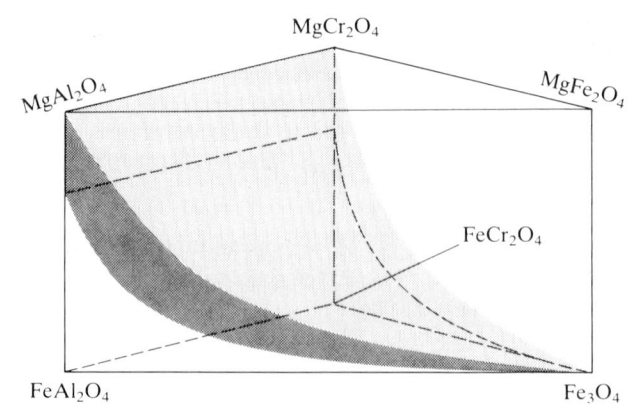

Transparent spinels have high relief ($n > 1.7$) in thin section and are isotropic. An octahedral habit, sometimes with twinning on $\{111\}$, helps to distinguish them from the garnets. Opaque spinels are isotropic; they differ in their reflectance values.

There is considerable variation in the chemical composition of natural spinels, and this leads to a variety of colours and degrees of opacity (Table 3.2). Magnetic (ferrimagnetic) spinels also exist, the best known being magnetite. The spinels can often be identified on the basis of their mode of occurrence, textural relations and associated phases, but chemical analysis is usually required for satisfactory identification.

Cassiterite SnO_2

Cassiterite may contain minor amounts of Fe, Nb, Ta, Ti, W or Si.

Crystals Cassiterite is tetragonal with $a:c = 1:0.672$. Crystals are usually short prisms $[001]$ with $\{110\}$ and $\{100\}$ prominent (Fig. 3.11). Faces in $[10\bar{1}]$ and $[001]$ are often striated. Twinning on $\{011\}$ is very common

Table 3.2 Spinels.

Spinel	Composition	Structural type	Colour/opacity
spinel	$MgAl_2O_4$	normal	$n = 1.719$, colourless or red, blue brown etc.
hercynite	$FeAl_2O_4$	normal	$n = 1.835$, dark green to black
magnetite	Fe_3O_4	inverse	opaque
maghemite	Fe_2O_3	inverse	opaque (metastable with respect to hematite)
ulvospinel	Fe_2TiO_4	inverse	opaque
chromite	$FeCr_2O_4$	normal	opaque, dark brown on edges
Two common intermediate varieties are:			
pleonaste	$(Mg,Fe)Al_2O_4$	normal	green to blue green
picotite	$(Fe,Mg)(Al,Cr)_2O_4$		

Figure 3.11 Cassiterite crystals.

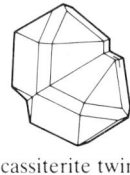

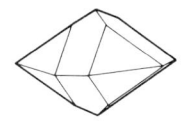

cassiterite twin on (011)

giving both contact and penetration twins. There is a poor $\{100\}$ cleavage.

Thin section
$n_o = 1.990–2.010$
$n_e = 2.093–2.100$
$\delta = 0.096–0.098$
Uniaxial +ve (crystals are length slow)
$D = 6.98–7.02 \quad H = 6–7$

COLOUR Colourless or slightly red or brown.
*PLEOCHROISM Occasionally present in coloured varieties with o pale colours and e dark yellow, brown, reddish.
CLEAVAGE $\{100\}$ and $\{010\}$ prismatic cleavages parallel to length of mineral.
*RELIEF Extremely high.
*BIREFRINGENCE Very high, but interference colours are often masked by the colour of mineral.
TWINNING Simple and repeated, common on $\{011\}$.
ZONING Variation in iron content leads to colour banding.
Polished section Cassiterite is grey, sometimes appearing slightly brownish. With $R_o = 11\%$ and $R_e = 13\%$, bireflectance is weak but usually visible in granular aggregates and twinned grains. Cassiterite is darker than sphalerite and only slightly brigher than gangue minerals. Anisotropy is distinct in greys but internal reflections, which are common and colourless to brown, often mask the anisotropy.

Cassiterite occurs as isolated prismatic to rounded crystals, geniculate (knee-like) twins or granular aggregates. Colloform aggregates containing colloidal hematite are known as 'wood tin'. Twinning is common and cleavage traces are often present. Zonation of iron content may be seen in crossed polars because Fe absorbs the light in the internal reflections. VHN = 1027–1075.

Occurrence Cassiterite is mainly found with wolframite, tourmaline, topaz, arsenopyrite, molybdenite, pyrrhotite and bismuthian minerals in high temperature hydrothermal veins, pegmatites, greisens, stockworks and disseminations associated with acid igneous rocks. It is found as a detrital heavy mineral in sediments (such as the commercial placer deposits of Malaysia) and in gossans over stanniferous sulphide deposits. Wood tin is found in the secondary oxidation zone.

Distinguishing features Compared with cassiterite, sphalerite is brighter, isotropic and softer; wolframite is slightly brighter and has fewer internal reflections; and rutile is brighter.

THE NON-SILICATES

Chromite $FeCr_2O_4$

Crystals Usually containing Mg and Al, chromite may also contain Zn, V, Mn.

Thin section Chromite is cubic and a member of the spinel group. Crystals are rare but occur as octahedra modified by $\{001\}$ faces. There is no cleavage. $D = 5.1$.

Chromite is opaque except in very thin grain margins which are brownish in colour.

Polished section Chromite is grey, sometimes appearing slightly brownish. $R = 12\%$ but varies with chemical composition. This reflectance value is significantly less than that of magnetite. High Fe and Cr values increase R but Al and Mg decrease R. Although cubic and usually isotropic, chromite sometimes shows weak anisotropy. Iron-poor chromite may have scarce reddish brown internal reflections.

Chromite occurs as rounded octahedral grains resembling droplets, interstitially in silicates, or as granular aggregates. It is an accessory mineral in most peridotites and derived serpentinites. Cataclastic texture is common. A zonation in reflectance related to chemical zonation may be observed. Marginal discoloration and alteration may occur. Inclusions of Fe + Ti + O phases, e.g. rutile, may be present. VHN = 1195–1210.

Occurrence Chromite is only abundant in certain mafic igneous rocks, especially large layered intrusions (e.g. the Bushveldt lopolith) as cumulates or possibly oxide–liquid segregations. It is found as podiform concentrations, possibly originally cumulates, in Alpine-type serpentinites and also as a detrital heavy mineral in sedimentary and metamorphic rocks. Chromite may occur as cores within magnetite grains. Iron-rich rims of chromites, commonly observed in serpentinites, are known as ferritchromit. The rims have a slightly higher reflectance than the chromite cores and are magnetic.

Distinguishing features Compared with chromite, magnetite is brighter. The two minerals are similar unless direct comparison of brightness can be made. However, remember that magnetite is magnetic!

Corundum Al_2O_3 trigonal c/a 1.364

$n_o = 1.768–1.772$
$n_e = 1.760–1.763$
$\delta = 0.008–0.009$
Uniaxial −ve (crystal hexagonal, rarely prismatic)
$D = 3.98–4.02$ *$H = 9$

COLOUR Colourless, but gem quality corundum is often coloured blue (sapphire) or red (ruby) in hand specimen.

PLEOCHROISM Normal corundum is not pleochroic but gem quality minerals are weakly pleochroic, particularly sapphire with e blue and o light blue.

HABIT Rarely euhedral, usually as small rounded crystals.

*CLEAVAGE	None; basal parting present.
*RELIEF	High (about the same as garnet).
ALTERATION	Corundum can alter to Al_2SiO_5 minerals during metamorphism, by addition of silica, or to muscovite if water and potassium are also available.
BIREFRINGENCE	Low, common interference colours are first order greys and whites.
*TWINNING	Lamellar twinning is commonly seen on $\{10\bar{1}1\}$. Simple twins can occur with $\{0001\}$ as the twin plane.
DISTINGUISHING FEATURES	Corundum has low birefringence, high relief, no cleavage and lamellar twinning. Apatite has lower relief and still lower birefringence.
*OCCURRENCE	Corundum occurs in silica-poor rocks such as nepheline–syenites, and other alkali igneous undersaturated rocks. It may occur in contact aureoles in thermally altered aluminous shales, and in aluminous xenoliths found within high temperature basic igneous plutonic and hypabyssal rocks. In these aluminous xenoliths, corundum is frequently found in association with spinel, orthopyroxene and cordierite. Corundum occurs in metamorphosed bauxite deposits, and also in emery deposits. It can occur as a detrital mineral in sediments.

Hematite Fe_2O_3

Hematite is often titaniferous, i.e. there is a hematite–ilmenite solid solution. See Section 3.4.

Crystals Hematite is hexagonal, $a:c = 1:1.3652$ and usually occurs as tabular crystals $\{0001\}$ often in subparallel growths (Fig. 3.12). Penetration twinning occurs on $\{0001\}$ and lamellar twinning on $\{10\bar{1}1\}$. There is no cleavage. $D = 5.2$.

Thin section Hematite is opaque but deep red in very thin plates. It is uniaxial $-$ve with absorption o $>$ e.

Polished section Hematite is light grey and only weakly bireflecting, with $R_o = 30\%$ and $R_e = 25\%$. It is much brighter than magnetite and ilmenite. Anisotropy is strong in bluish and brownish greys. The deep red internal reflections are scarce except in very thin plates. Hematite coatings give a red colouration to internal reflections of transparent grains such as quartz.

Hematite occurs as idiomorphic tabular crystals and fibrous radiating aggregates. It is also found as microcrystalline colloform masses. It is often intergrown with other Fe + Ti + O minerals and occurs as lamellae in ilmenite. Hematite may contain lamellae of ilmenite or rutile. Lamellar twinning is common and a pseudo-cleavage consisting of elongate pits may be present. VHN = 920–1062.

Figure 3.12 Typical hematite crystals.

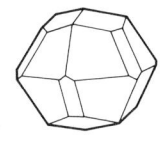

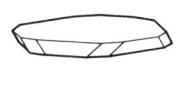

THE NON-SILICATES

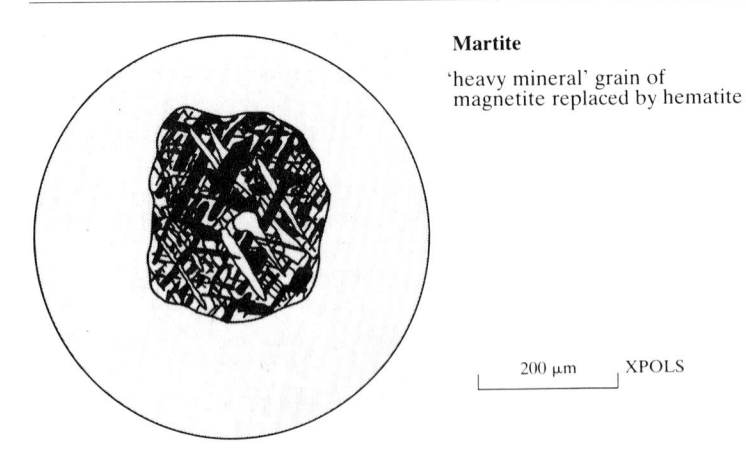

Martite

'heavy mineral' grain of magnetite replaced by hematite

200 μm   XPOLS

Occurrence Hematite is found with other Fe-Ti-O minerals in igneous and metamorphic rocks as well as sedimentary rocks, especially banded iron formation. Hematite in veins can be primary but it frequently forms by oxidation of other primary iron-bearing minerals, for example in gossans.

Distinguishing features Compared with hematite, stibnite has a distinct bireflectance, is softer and has a good cleavage; ilmenite is pinkish and darker; and cinnabar has abundant internal reflections and is softer.

Note Martite is magnetite pseudomorphed by an intergrowth of hematite.

Ilmenite $FeTiO_3$

Ilmenite may contain Mn or Mg, the magnesian end member being geikielite and the manganiferous end member being pyrophanite. It may also contain Fe^{3+}, which represents a solid solution towards hematite Fe_2O_3.

Crystals Ilmenite is trigonal, $a:c = 1:1.3846$, and occurs as tabular $\{0001\}$ crystals. Twinning occurs on $\{0001\}$ and multiple twinning on $\{10\bar{1}1\}$. There is no cleavage but there is a parting parallel to $\{10\bar{1}1\}$. $D = 4.7$.

Thin section In very thin flakes ilmenite is red, uniaxial $-$ve.

Polished section Ilmenite is slightly pinkish or brownish light grey with a weak pleochroism. $R_o = 21\%$, which is similar to magnetite, and $R_e = 18\%$ Anisotropy is moderate but only distinct in some orientations; tints are greenish, bluish and brownish greys.

Ilmenite is sometimes idiomorphic but is usually intergrown with other Fe-Ti-O minerals. It often contains lamellar inclusions of hematite or other Fe-Ti-O minerals. Occasionally lamellar twins may be present. VHN = 519–703, varying with chemical composition.

Occurrence Ilmenite is found with other Fe-Ti-O minerals in igneous rocks (especially of mafic composition) and metamorphic rocks, and also (but rarely) in veins and pegmatites. Detrital ilmenite is usually altered to leucoxene which is enriched in TiO_2. It occurs in heavy mineral con-

OXIDES

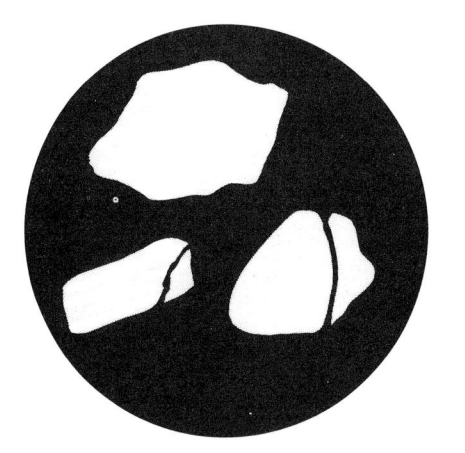

Ilmenite

grains of ilmenite with exsolved hematite

500 μm PPL

centrates. Magnesium-rich ilmenites occur in kimberlites but also in contact metamorphosed rocks.

Distinguishing features Compared with ilmenite, magnetite is slightly brighter and usually bluish grey in direct comparison, isotropic and strongly magnetic, and rutile shows abundant internal reflections.

Magnetite Fe_3O_4

Magnetite often contains Ti, Cr or Mn. Titaniferous magnetite often contains ulvospinel Fe_2TiO_4 in solid solution.

Crystals Magnetite is an inverse spinel. It is cubic, commonly occurring as octahedra and combinations of the octahedron and rhombic dodecahedron. Twinning is common on $\{111\}$, the usual spinel twin. $D = 5.2$.

Polished section Magnetite is grey, sometimes with a brownish or pinkish tint indicative of titanium. (Ulvospinel is brownish grey). $R = 21\%$, making magnetite much darker than pyrite and hematite. Magnetite is isotropic with good extinction.

Magnetite

a grain of magnetite with exsolved ilmenite

200 μm PPL

THE NON-SILICATES

It is often found as idiomorphic octahedral sections, but also as skeletal grains or granular aggregates. Lamellae of hematite are often in a triangular pattern. Lamellae and blebs of ilmenite in a fine 'frosty' texture of ulvospinel in magnetite represents slowly cooled titaniferous magnetite. Also, exsolved lamellae and blebs of dark grey spinels may be present. VHN = 530–599.

Occurrence Magnetite is found usually with other Fe-Ti-O minerals in igneous and metamorphic rocks and skarns. It also occurs as a heavy mineral in sediments and sedimentary rocks and in high temperature hydrothermal veins with sulphides. It represents reducing conditions relative to hematite.

Distinguishing features Compared with magnetite, ilmenite is similar but often pinker and anisotropic; sphalerite is softer, usually has internal reflections and occurs in a different association; and chromite is very similar in isolation but is darker and may show internal reflections. A magnetised needle may be used to confirm the magnetism of magnetite grains in polished section.

Rutile TiO_2

Rutile may contain some Fe or Nb. The polymorphs anatase and brookite are almost identical to rutile in polished section.

Crystals Rutile is tetragonal, $a:c = 1:0.6442$. Crystals are commonly prismatic, often slender to acicular (Fig. 3.13). Twinning on $\{011\}$ is common and is often repeated or geniculated. There is a distinct cleavage on $\{110\}$ as well as a cleavage on $\{100\}$.

Thin section $n_o = 2.605$–2.613
$n_e = 2.899$–2.901
$\delta = 0.286$–0.296
Uniaxial +ve (crystals are length slow)
$D = 4.23$–5.5 $H = 6$–$6\frac{1}{2}$

*COLOUR Reddish brown or yellowish (sometimes opaque).
PLEOCHROISM Common, although sometimes weak with o pale yellow, and e pale brown dark brown, red.

rutile crystal

rutile: twinning on (011)

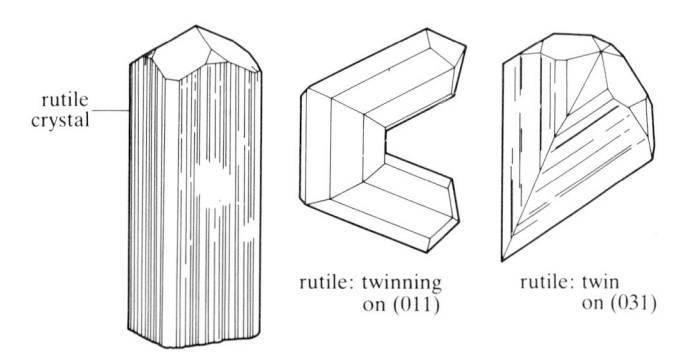

rutile: twin on (031)

Figure 3.13 Rutile crystal and twinning.

OXIDES

HABIT
: Small, acicular prisms common.

CLEAVAGE
: $\{110\}$ and $\{100\}$ prismatic cleavages good.

*RELIEF
: Exceptionally high.

*BIREFRINGENCE
: Extremely high, interference colours usually masked by mineral colour.

TWINNING
: Common in a number of different planes.

Polished section
: Rutile is light grey with a slight bluish tint. $R_o = 20\%$ and $R_e = 23\%$. The bireflectance is weak but usually distinct. Rutile has about the same brightness as magnetite. It is strongly anisotropic in greys but anisotropy is often masked by abundant bright colourless yellow to brown internal reflections. In iron-rich varieties, internal reflections are less abundant and reddish.

    Rutile occurs as prismatic to acicular isolated crystals or as aggregates of crystals and in spongy porphyroblasts. It usually occurs as small grains. Multiple and simple twins are common. VHN = 1074–1210.

Occurrence
: Rutile is associated with other Fe-Ti-O phases in pegmatites, igneous and metamorphic rocks. It is a heavy mineral in sediments. It is often produced from ilmenite on wall rock alteration by hydrothermal solutions, e.g. greisenisation. Rutile occurs within quartz crystals as long thread-like crystals.

Distinguishing features
: Compared with rutile, hematite is whiter and brighter and rarely shows internal reflections; ilmenite is slightly pinkish and does not show internal reflections; and cassiterite tends to be more equant and is darker.

Notes
: The low temperature TiO_2 polymorphs anatase (uniaxial −ve) and brookite (orthorhombic, small $2V$ +ve) have a similar occurrence to rutile. In polished section anatase lacks twinning and is only very weakly anisotropic. Anatase occurs associated with clay minerals in sedimentary rocks.

Spinel $MgAl_2O_4$

The two common aluminous spinels are spinel $MgAl_2O_4$ and hercynite $FeAl_2O_4$. The general formula of aluminous spinels is $M^{2+}M_2^{3+}O_4$, with M^{2+} = Mg, Fe, Mn, Zn and M^{3+} = Al. There is extensive solid solution including $Al^{3+} \rightleftharpoons Fe^{3+}, Cr^{3+}$.

Crystals
: Spinel is cubic and usually of octahedral habit (Fig. 3.14). Twinning on $\{111\}$ may be repeated. There is a poor $\{111\}$ cleavage. $D = 3.55$ ($D = 4.40$ hercynite).

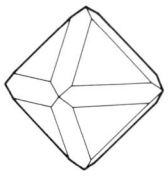

Figure 3.14 Spinel crystal and twinning.

spinel crystal

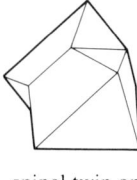

spinel twin on (111)

THE NON-SILICATES

Thin section Spinel is of variable colour and opacity. Mg-rich spinel is transparent and isotropic.

Polished section Spinel is grey with $R = 8\%$, making it only slightly brighter than associated silicates. It is isotropic. Internal reflections vary in abundance depending on composition.

Spinel is often idiomorphic or rounded octahedral. It may contain inclusions of magnetite or ilmenite. VHN = 861–1650 spinel; 1402–1561 hercynite.

Occurrence Spinel occurs as exsolved blebs or lamellae in magnetite. It is found in basic igneous rocks and contact metamorphic and metasomatic aluminous (or Si-deficient) rocks. It is also found as a heavy mineral in placer deposits. Unlike spinel, hercynite is stable in the presence of free silica.

Uraninite UO_2

Natural uraninite is often oxidised to some extent to pitchblende UO_{2-3}. The U is often replaced by Th or Ce.

Crystals Uraninite is cubic and usually occurs as octahedra, cubes or dodecahedra. Twinning on $\{111\}$ is rare. There is no cleavage. $D \approx 9.0$.

Thin section Uranium oxides often appear as opaque rounded aggregates altered along fractures. In thin splinters a green to brown colour may be obtained. Associated minerals may be darkened due to radiation damage.

Polished section Uraninite is grey with $R = 17\%$, similar to sphalerite. It is cubic and isotropic. Pitchblende is similar but slightly darker, with $R = 16\%$. Scarce brown internal reflections may be observed in these minerals.

Uranium oxides commonly occur as spherical or botryoidal masses. Uraninite is well crystallised but pitchblende varies in crystallinity and non-stoichiometry and tends to polish poorly. Composition zoning results in slight brightness and hardness changes. Shrinkage cracks occur in pitchblende. VHN = 782–839 uraninite; 673–803 pitchblende.

Pitchblende

note the 'patchiness' of the brightness due to variation in oxidation: shrinkage cracks radiate from the centre of spheroids

500 μm   PPL

HALIDES

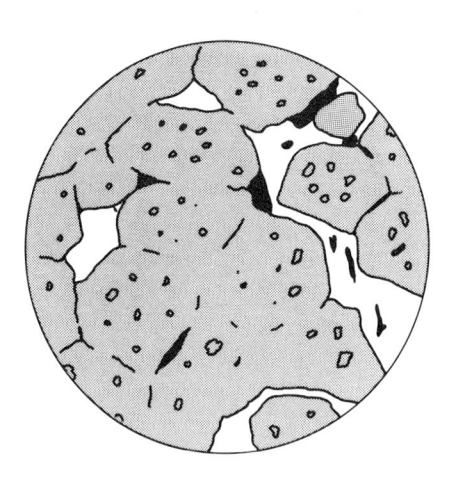

Thucolite

carbon (dark grey) with inclusions of uraninite (dark grey): pyrite and gold (both white) are interstitial – typical of cross sections of Witwatersrand columnar thucolite

200 μm   PPL

Occurrence Uranium oxides are found in high temperature pegmatitic to low temperature hydrothermal vein and replacement deposits. There is an association with Ni + Co + Ag + Bi mineralisation, with acid igneous rocks and with organic material in sedimentary rocks. Detrital uraninite is found in placer deposits with gold.

Distinguishing features Compared with uraninite, magnetite is similar but is magnetic. The uranium oxides are radioactive!

Notes Oxidation of primary uranium oxides often results in distinctive bright yellow or green secondary uraniferous minerals.

Thucolite is fragmental uraninite in polymerised carbonaceous material.

3.5 Halides

Halides are ionic minerals that consist essentially of metallic cations and halogen anions. The common examples are normal anhydrous halides, which have simple chemical compositions and structures, but there is a host of related oxyhalides, hydroxyhalides and complex-containing halides. One of the aluminofluorides, cryolite Na_3AlF_6, is worthy of mention, but the others are of little significance as rock-forming minerals. The structures of fluorite CaF_2 and halite (rock salt) NaCl are illustrated in Figure 3.15. Sylvite KCl has the halite structure.

The simple halides are typical ionic comounds, there being a large difference in the ionisation potentials of the metal and halogen atoms; they are typical transparent insulators. Both halite and fluorite have low refractive indices and are therefore quite dark (i.e. they have a small reflectance) in polished section.

THE NON-SILICATES

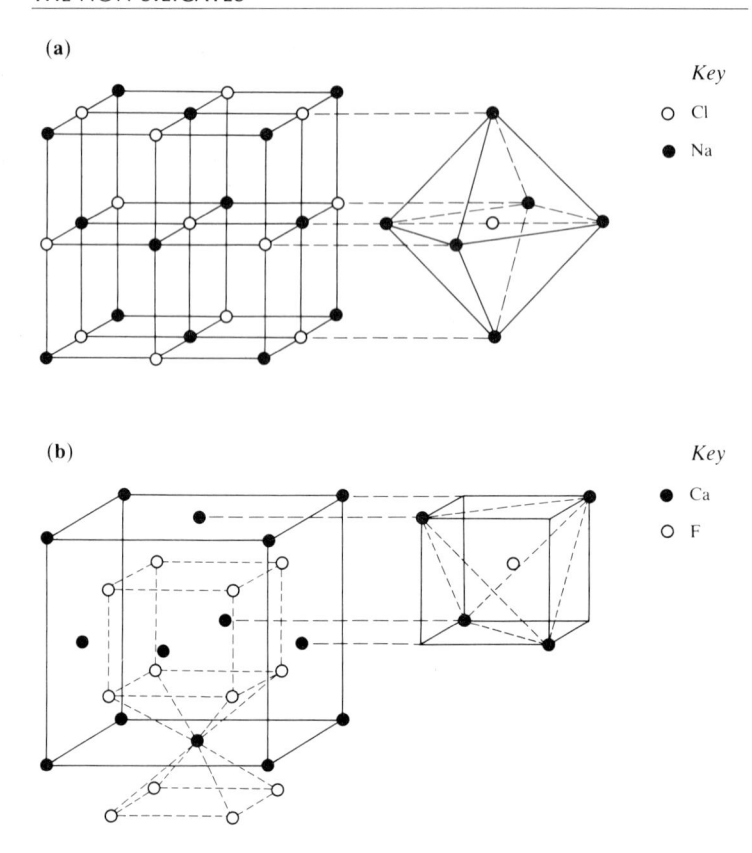

Figure 3.15 (a) Crystal structure of sodium chloride, showing (right) the octahedral arrangement of six sodium ions around one chloride ion. (b) Fluorite structure showing (right) the tetrahedral arrangement of four calcium ions around one fluoride ion, and (below) the cubical arrangement of eight fluoride ions around one calcium ion.

Fluorite CaF_2 cubic

n = 1.433 – 1.435 (variation due to substitution of Y for Ca)
D = 3.18 H = 4

COLOUR   Colourless, very pale green, pale blue, yellow or violet.
HABIT   Aggregates of crystals often with perfect cubic $\{100\}$ form.
CLEAVAGE $\{111\}$ perfect, giving a triangular pattern.
*RELIEF   Moderate, but note that n is less than 1.54.
TWINNING   Interpenetrant on $\{111\}$, but not seen in thin section.
*OCCURRENCE   A late stage mineral in granites and other acid rocks; common in greisen. In pegmatites, and many alkaline igneous rocks such as nepheline–syenites, fluorite crystallises at low temperatures (around 500 °C). In

late stage pneumatolytic deposits fluorite occurs with cassiterite, topaz, apatite and lepidolite, whereas in hydrothermal veins fluorite occurs with calcite, quartz, barite and sulphides.

Fluorite is occasionally found as the cementing matrix in sandstones and may occur in geodes within limestones. Blue John is coarse nodular purple fluorite with a concentric layered structure.

Halite (rock salt) NaCl   cubic
$n = 1.544$
$D = 2.16 \quad H = 2\frac{1}{2}$

Halite is colourless rock salt with a perfect $\{100\}$ cubic cleavage. It occurs in salt domes and in evaporites, where it is a late precipitating salt.

Note: Special sectioning techniques are needed to preserve this mineral in thin sections.

3.6 Hydroxides

Brucite Mg(OH)$_2$ trigonal, c/a 1.521
$n_o = 1.560–1.590$
$n_e = 1.580–1.600$
$\delta = 0.012–0.020$
Uniaxial +ve (length fast)
$D = 2.4 \quad H = 2\frac{1}{2}$

COLOUR Colourless.
HABIT Occurs as fine aggregates, or fibrous whorls, in metamorphosed impure limestones.
*CLEAVAGE Perfect basal $\{0001\}$.
RELIEF Low, just greater than 1.54.
ALTERATION Brucite forms from periclase MgO by addition of H$_2$O during thermal metamorphism. It alters to hydromagnesite readily by reaction with carbon dioxide:

$$5Mg(OH)_2 + 4CO_2 \rightarrow Mg_5(OH)_2(CO_3)_4 \cdot 4H_2O$$
$$\text{hydromagnesite}$$

*BIREFRINGENCE Low, first order colours, but often shows anomalous interference colours (deep blue) rather similar to chlorite.
TWINNING None.
*OCCURRENCE Brucite occurs in thermally metamorphosed dolomites, and dolomitic limestones. It can occur in low temperature hydrothermal veins, associated with serpentinites and chlorite schists.

THE NON-SILICATES

DISTINGUISHING FEATURES
Brucite has anomalous birefringence and is uniaxial positive. Micas and talc have higher birefringence and are optically negative, as is gypsum. Apatite and serpentine are very similar optically, but serpentine is usually greenish and both are *always* length slow.

Limonite
Goethite
Lepidocrocite

$FeO.OH.nH_2O$

Limonite is brown earthy material consisting of goethite ± lepidocrocite with absorbed water.

Crystals Goethite α-$HFeO_2$ is orthorhombic, $a:b:c = 0.4593:1:0.3034$. Lepidocrocite γ-FeO.OH is orthorhombic, $a:b:c = 0.309:1:0.245$. Both minerals occur as flakes or blades flattened (010) or as fibres elongated [100]. There is perfect cleavage on {100}, {010} and {001}. $D = 4.28$ (goethite); $D = 4.0$ (lepidocrocite).

Thin section Poorly crystallised limonite appears reddish brown and isotropic. Goethite is yellowish to brownish, pleochroic with absorption $\alpha < \beta < \gamma$, and has a small $2V$, −ve. Lepidocrocite is yellow to brownish red, strongly pleochroic with absorption $\alpha < \beta < \gamma$, and has $2V = 83°$, −ve.

Polished section Poorly crystallised limonite is bluish grey with $R = 16$–19%; anisotropy is strong in bluish greys, and deep red to brown internal reflections are typical. Goethite is grey with $R \approx 17\%$; anisotropy is distinct in shades of grey. Lepidocrocite is grey with $R \approx 10$–19%; anisotropy is very strong in slightly bluish light greys. Internal reflections are deep red to brown in both minerals.

Limonite is often inhomogeneous, varying in colour or porosity. Goethite is usually colloform and botryoidal whereas lepidocrocite is usually better crystallised. VHN: 772–824 goethite; 690–782 lepidocrocite.

Limonite

'heavy mineral' grain of pyrite (white) oxidised and replaced by limonite (grey)

300 μm   PPL

Occurrence Limonite is very common as a weathering product after iron-bearing minerals, especially iron carbonates and iron sulphides. It is associated with other hydroxides and oxides in various types of gossans.

Distinguishing features Compared with limonite, hematite is brighter, harder and has only scarce internal reflections, and sphalerite is isotropic and usually differs texturally.

3.7 Sulphates

Anhydrite $CaSO_4$ orthorhombic
$0.893:1:1.001$

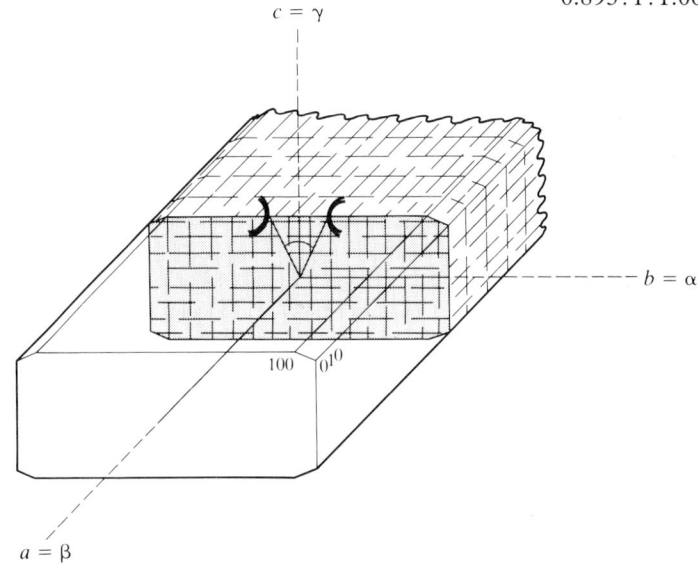

$c = \gamma$

$b = \alpha$

100 010

$a = \beta$

$n_\alpha = 1.569–1.574$
$n_\beta = 1.574–1.579$
$n_\gamma = 1.609–1.618$
$\delta = 0.04$
$2V_\gamma = 42°–44°$ +ve

OAP is parallel to (100). Crystal is elongated along the a axis and can be length fast or slow

$D = 2.9–3.0 \quad H = 3–3½$

COLOUR Colourless.
HABIT Prismatic crystals with aggregates common.
CLEAVAGE {010} perfect, {100} and {001} good.
RELIEF Low to moderate.
ALTERATION Just as gypsum can dehydrate to anhydrite, so anhydrite can react with water to form gypsum.

THE NON-SILICATES

*BIREFRINGENCE High (much greater than gypsum) with third order colours.
*INTERFERENCE FIGURE Seen on an elongate prism section, a Bx_a figure is just bigger than field of view and positive.
*EXTINCTION Straight on all cleavages and prism edges.
TWINNING Repeated on $\{011\}$.
*OCCURRENCE Similar to gypsum, being found in evaporite deposits. Anhydrite may form by hydrothermal alteration of limestones or dolomites.

Barite (baryte or **barytes)** $BaSO_4$ orthorhombic
Celestite (celestine) $SrSO_4$ $0.815:1:1.314$

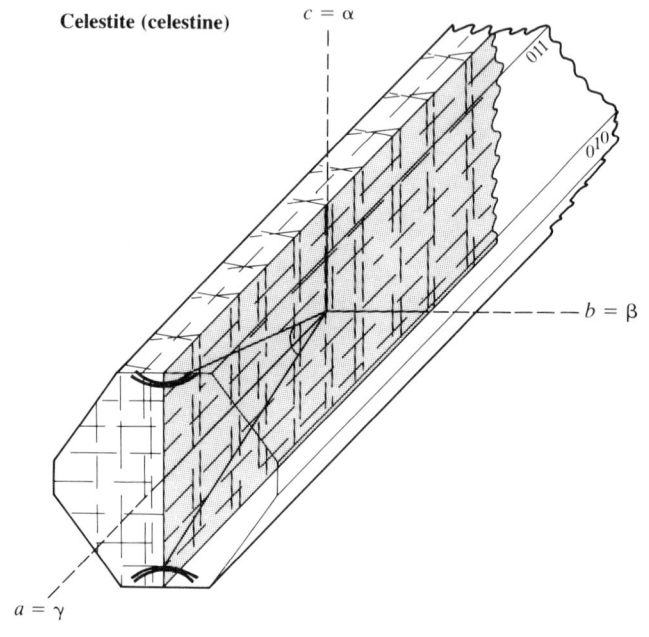

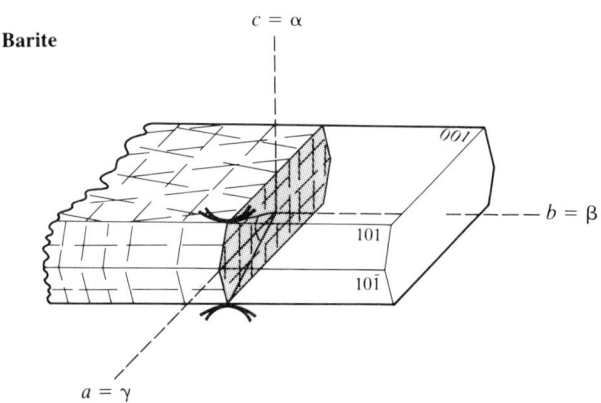

Celestite (celestine)

Barite

$n_\alpha = 1.636$
$n_\beta = 1.637$
$n_\gamma = 1.647$
$\delta = 0.012$
$2V_\gamma = 37°$ +ve (length slow)
OAP is parallel to (010). Crystal is elongated parallel to the a axis
$D = 4.5 \quad H = 2½–3½$ (barite)
$D = 4.0 \quad H = 2½–3½$ (celestite)

COLOUR Colourless.
HABIT Subhedral clusters of prismatic crystals common.
CLEAVAGE Basal cleavage {001} perfect, {210} and {010} cleavages present.
RELIEF Moderate.
BIREFRINGENCE Low, first order yellows, but mottled colours are common.
*INTERFERENCE FIGURE Bx_a figure with small $2V$ seen on (100) section, i.e. section with two cleavages.
EXTINCTION Straight on cleavage trace or prism edge.
TWINNING Lamellar twinning present on {110}.
*OCCURRENCE Barite occurs as a gangue mineral in ore-bearing hydrothermal veins, in association with fluorite, calcite and quartz. Barite occurs as stratiform deposits of synsedimentary exhalative origin in sedimentary and metamorphic terrains.

Celestite $SrSO_4$ is similar to barite optically and occurs in dolomites, in evaporite deposits, and rarely in hydrothermal veins. There is probably a complete solid solution series from barite to celestite.

THE NON-SILICATES

Gypsum $CaSO_4.2H_2O$
monoclinic
$0.6899:1:0.4214, \beta = 99°18'$

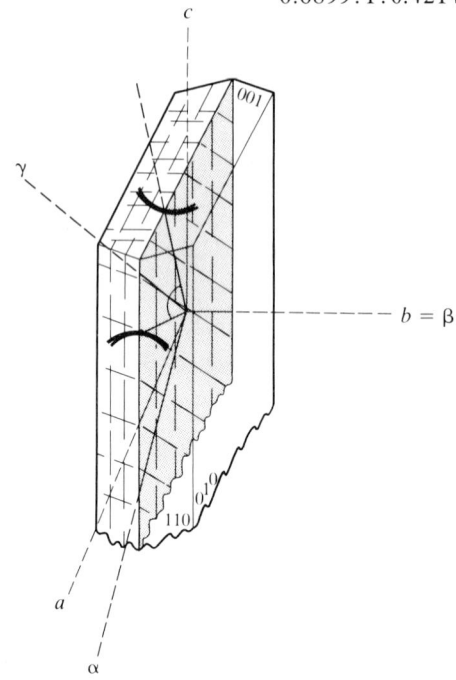

n_α = 1.519–1.521
n_β = 1.523–1.526
n_γ = 1.529–1.531
δ = 0.01
$2V_\gamma \approx 58°$ +ve
OAP is parallel to (010)
D = 2.30–2.37 H = 2

COLOUR — Colourless.
HABIT — Anhedral crystals occur usually in aggregate masses.
CLEAVAGE — {010} perfect, {100} and {011} good.
RELIEF — Low, always less than CB.
ALTERATION — With increase in temperature gypsum changes to anhydrite thus (about 200 °C):

$$\int CaSO_4.2H_2O \rightarrow CaSO_4 + 2H_2O$$

BIREFRINGENCE — Low, interference colours are first order whites.
INTERFERENCE FIGURE — Bx_a figure is seen on a thin prismatic section but $2V$ is larger than field of view; thus the sign is best determined by looking at an optic axis figure.
EXTINCTION — Straight on the {010} cleavage.
TWINNING — Common on {100}, repeated twinning usually seen.

*OCCURRENCE Gypsum is mainly found in sedimentary rocks, especially in evaporitic sequences.

Calcium sulphate can occur as either gypsum or anhydrite. Anhydrite may be formed by the dehydration of primary gypsum. In desert regions calcium sulphate is dissolved in percolating ground waters, which can be drawn to the surface and deposit gypsum as 'desert roses' during very dry spells.

Gypsum can form in fissures in shales and other argillaceous rocks by the action of acid ground waters (sulphuric acid in solution), reacting with calcium either from limestone nodules within the argillaceous rocks or from intercalated limestone beds.

3.8 Phosphate

Apatite $Ca_5(PO_4)_3(OH,F,Cl)$ hexagonal, c/a 0.73

n_o = 1.623–1.667
n_e = 1.624–1.666
δ = 0.001–0.007
Uniaxial −ve (a prism section is length fast)
D = 3.1–3.35 H = 5

COLOUR Colourless.
*HABIT Small prismatic crystals with hexagonal cross section, often found with ferromagnesian minerals in rocks, particularly amphiboles and micas.
CLEAVAGE Good basal $\{0001\}$ cleavage, imperfect prismatic cleavage $\{1010\}$.
RELIEF Moderate.
*BIREFRINGENCE Very low, maximum interference colours are grey.
OCCURRENCE Important accessory mineral in igneous rocks, especially acidic plutonic rocks, granite pegmatites and vein rocks, but common in diorites and gabbros also.

Apatite is common in metamorphic rocks, especially chlorite schists and amphibole-bearing schists and gneisses.

Apatite occurs as a detrital mineral in sedimentary rocks. Sedimentary phosphatic deposits commonly contain a cryptocrystalline phosphatic mineral called 'collophane', a term used if apatite cannot be positively identified.

3.9 Tungstate

Wolframite (wolfram) $(Fe,Mn)WO_4$

The iron end member is called ferberite and the manganese end member huebnerite.

Crystals Wolframite is monoclinic, $a:b:c$ = 0.8255:1:0.8664, β = 90°29'. It is usually prismatic $[001]$. Simple twinning is common and takes place on

THE NON-SILICATES

$\{100\}$ and $\{023\}$. There is a perfect $\{010\}$ cleavage and a parting on $\{100\}$ and $\{101\}$.

Thin section $n_\alpha = 2.150–2.269$
$n_\beta = 2.195–2.328$
$n_\gamma = 2.283–2.444$
$\delta = 0.133–0.175$
$2V_\gamma = 60°–70°$ +ve
OAP is perpendicular to (010)
$D = 7.18–7.61$ $H = 5–5½$.

COLOUR Transparency decreases with increase in Fe content. Colour banding is due to a variation in Fe:Mn ratio. Iron-rich wolframite is brownish red to dark green.

PLEOCHROISM Common, with α red, brown, yellow, β pale green to yellowish brown, γ red, green, dark brown.

HABIT Elongate prismatic often occurring as thin flat crystals.

CLEAVAGE $\{010\}$ perfect.

RELIEF Extremely high.

*BIREFRINGENCE Extremely high, but colours hidden by mineral colour.

INTERFERENCE FIGURE Seen on section perpendicular to (010), $2V$ large, but difficult to determine sign since high dispersion of mineral will make determination colours difficult to see.

EXTINCTION Oblique extinction with $\gamma\char`\^cl = 17°$ to $27°$.

Polished section Wolframite is slightly brownish grey. With $R \approx 17\%$ it is slightly brighter than cassiterite. Bireflectance is weak. Anisotropy is moderate and distinct in bluish greys. Extinction is oblique. Reddish brown internal reflections are common.

Wolframite occurs as idiomorphic tabular or bladed crystals with simple twinning. Zoning is enhanced by weathering. Cleavage traces may be observed. VHN = 357–394.

Occurrence Wolframite is found in high temperature hydrothermal veins and pegmatites usually associated with quartz and Sn, Au and Bi minerals. It is associated with granitic rocks and greisenisation. It is also found in placers with cassiterite. Scheelite $CaWO_4$ is a common associate and may replace wolframite.

Distinguishing features Compared with wolframite, cassiterite is darker and has more abundant internal reflections, and sphalerite is isotropic and is often associated with chalcopyrite.

3.10 Arsenide

Niccolite NiAs

Nickeline is the name recommended by the International Mineralogical Association. Niccolite may contain some Fe or Co.

NATIVE ELEMENTS

Crystals Niccolite is hexagonal, $a:c = 1:1.3972$. Crystals are rare. It is usually massive, reniform with columnar structure. Repeated twinning occurs on $\{10\bar{1}1\}$. There is no cleavage. $D = 7.8$.

Polished section Niccolite is pinkish or orange white with a pronounced pleochroism, with $R_o = 58\%$ (lighter, orange or yellowish) and $R_e = 52\%$ (darker, pinkish). The reflectance is similar to pyrite. Anisotropy is very strong, the tints being bright bluish and greenish greys.

Niccolite usually occurs in xenomorphic or concentric, botryoidal masses with other Co + Ni + As + S minerals. Grains are often cataclased. Growth zonation is common, and botryoidal masses often contain radiating intergrown irregular lamellae. VHN = 328–455.

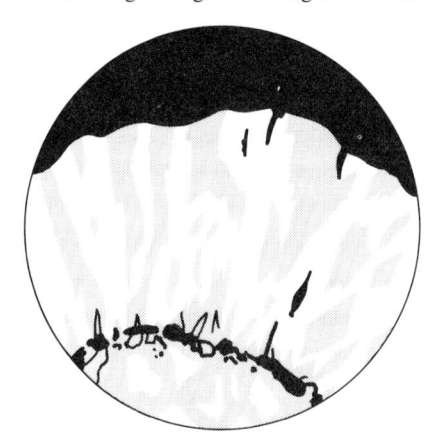

Niccolite

radiating intergrowth of niccolite in different crystallographic orientations

500 μm   PPL

Occurrence Niccolite occurs in Ni + Co + Ag + As + U deposits which are probably low temperature hydrothermal veins and replacements. Such deposits are often associated with basic igneous rocks and organic-rich sedimentary rocks.

Distinguishing features Compared with niccolite, marcasite is whiter, and arsenopyrite is whiter and has a weaker anisotropy.

Note Niccolite alters to green annabergite.

3.11 Native elements

Copper Cu

Copper may contain As, Ag or Bi.

Crystals Copper is cubic. $D = 8.95$.

Polished section Copper is bright metallic pink but tarnishes and darkens rapidly. $R = 81\%$. It is isotropic, but with incomplete extinction and fine scratches may cause false anisotropy.

Occurrence Copper occurs as small flakes, granular aggregates, porous masses or dendrites. Zonal texture is not uncommon, and lamellar twinning may be revealed by etching. VHN = 120–143.

THE NON-SILICATES

It is associated with cuprite Cu_2O and Cu + Fe + S minerals, often in deposits associated with basic extrusives. Copper is common in the oxidation zone, where it results from the oxidation of copper sulphides.

Distinguishing features   Compared with copper, gold is brighter and coloured yellow or white.

Gold   Au

Gold may contain Ag, Cu, Pd or Rh.

Crystals   Gold is cubic and occurs as cubic, dodecahedral or octahedral crystals, but repeated twinning on $\{111\}$ often gives reticulated and dendritic aggregates. $D = 19.3$.

Polished section   Gold is bright yellow. Argentiferous gold is whiter and cupriferous gold is pinker. $R = 74\%$, making gold much brighter than pyrite and chalcopyrite. It is isotropic but with incomplete extinction when a greenish colour is observed. Gold does not tarnish, but large grains scratch easily and may be difficult to polish.

Gold occurs as irregular grains, blebs or veinlets, often in sulphides (e.g. pyrite, arsenopyrite). The various varieties of gold are often intergrown with each other or with Au + Bi + Te and Sb + As-containing minerals. Gold occurs as very fine coatings which can easily be lost on polishing. VHN = 50–52.

Occurrence   Gold is found in hydrothermal deposits, often associated with igneous rocks; in placer deposits, where it appears to be chemically mobile, resulting in nugget growth; and in auriferous quartz veins. It seems to be present throughout the temperature range of vein mineralisation. Gold often occurs as very small grains, even in economic gold deposits.

Distinguishing features   Compared with gold, chalcopyrite is less yellow, darker and weakly anisotropic.

Notes   Electrum (Au,Ag) contains 30 to 45% Ag. It is brighter ($R \approx 83\%$) and softer (VHN = 34–44) than pure gold.

Graphite   C

Crystals   Graphite is hexagonal, $a:c = 1:1.27522$. The layered structure results in a perfect $\{0001\}$ cleavage. Crystals are hexagonal tablets $\{0001\}$. $D = 2.1$.

Thin section   In very thin flakes graphite is deep blue and uniaxial −ve.

Polished section   Graphite is brownish grey with a marked pleochroism from $R_o = 16\%$ (grey) to $R_e = 6\%$ (dark brownish grey). It appears slightly brighter than gangue minerals. The anisotropy is strong in yellowish greys. Extinction is parallel to the cleavage (corresponding to the grey of R_o in PPL) but deformation commonly results in undulose extinction.

Graphite occurs as flakes, tabular crystals, aggregates or botryoidal masses. Flakes are sometimes very long and broken or buckled. The cleavage is usually evident and often deformed. In fact graphite is rather difficult to polish, and surfaces of large grains are often damaged. VHN = 12.

NATIVE ELEMENTS

Graphite

flakes of graphite showing 'buckled' cleavage traces

200 μm   XPOLS

Occurrence Graphite is common in metasediments, where it forms from organic material; when abundant, a graphitic schist results. Such graphite is indicative of reducing conditions, and pyrite is usually also present. Graphite also occurs in vein-like deposits and large masses, some of which are of uncertain origin.

Distinguishing feaures Compared with graphite, molybdenite is texturally similar but much brighter.

Note Small flakes of graphite in metamorphic rocks are much more evident using oil immersion.

Silver Ag

Silver may contain minor amounts of Au, Hg, As, Sb, Pt, Ni, Pb or Fe.

Crystals Silver is cubic. $D = 10.5$.

Polished section Silver is white but it soon tarnishes. With $R \approx 95\%$ it is much brighter than the common ore minerals. It is isotropic, but false anisotropy may result from fine polishing scratches.

Silver occurs in dendritic or irregular masses and as inclusions, often in silver-bearing sulphides or sulphur-poor minerals. VHN = 46–118.

Occurrence Silver is found with Co + Ni + Fe arsenides, usually associated with basic igneous rocks. It also occurs in the oxidised zones of galena-bearing veins. Many veins recorded as silver veins are in fact argentiferous galena veins, the silver being produced as a by-product of lead recovery. Silver is associated with native copper and it is often associated with carbonate.

4 Transmitted-light crystallography

4.1 Polarised light: an introduction

Light is an electromagnetic vibration but, for the purpose of transmitted- (and reflected) light microscopy, light can be considered as being simply the transfer of energy by vibrating 'particles' along a path from the source to the observer. White light consists of many rays, ranging in wavelength from 380 to 770 nm through the visible spectrum.

However, it is simpler to consider the idealised case of a single ray of monochromatic light, that is light of a single wavelength (Fig. 4.1a). The wave is generated by vibration of particles (e.g. A) lying along the path of the ray. If the light is non-polarised, the particles vibrate at random in a *plane* normal to the direction of the ray. If the light is linearly (or plane) polarised by means of a polarising filter, then the particles simply vibrate up and down along the *line* xy.

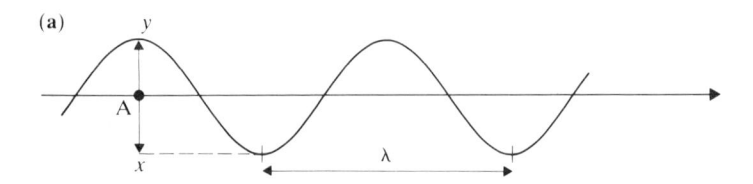

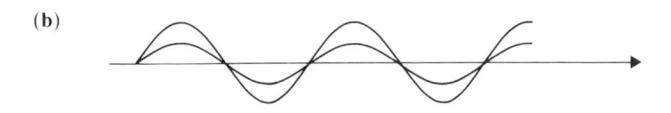

Figure 4.1 (a) Monochromatic light. (b) Two waves of the same wavelength, but different intensity, in phase. (c) Two waves of the same wavelength, but different intensity, out of phase.

Light of the same wavelength and the same or different intensity may be in phase as illustrated in Figure 4.1b, or out of phase as shown in Figure 4.1c. The path difference may be measured as a fraction of the wavelength.

If two coherent rays (originating at the same instant from the same source) which are exactly in phase are combined, they are added together and intensity is enhanced. If the rays are slightly out of phase, the enhancement is reduced. If the rays have the same amplitude and have a path difference of half a wavelength, the vibration will be cancelled and amplitude will be zero.

In transmitted-light microscopy, linearly polarised white light travels up the microscope axis, which is normal to the plane of the thin section. On entering an anisotropic (see Section 4.3) crystalline substance rotated from the extinction position, the light can be considered to be separated into two components which travel with different velocities through the crystal. On leaving the crystal the two components may be out of phase and the path difference will vary for different wavelengths of light. This complexity in the light leaving the crystal is only apparent when the analyser is inserted and interference colours are generated (see Section 4.6).

4.2 Refractive index

The refractive index of a medium (RI or n) is defined as the ratio of the velocity of light in the medium to that *in vacuo*. Refractive index varies with wavelength but the variation is usually small for transparent minerals, so single 'white light' refractive indices are usually used.

If V_1 and V_2 are the velocities of light in two different media, and i is the angle of incidence of the light in one medium and r is its angle of refraction in the other, then (see Fig. 4.2):

$$\frac{V_1}{V_2} = \frac{bc}{b'c'} = \frac{b'c \sin i}{b'c \sin r} = \frac{\sin i}{\sin r} = \text{refractive index}$$

The refractive index of a medium is inversely proportional to the velocity of light (for a specific wavelength) through the medium, i.e. $RI \propto 1/V$.

4.3 Isotropy

Isotropic crystals transmit light with equal velocity in all directions. A ray velocity surface represents the surface composed of all points reached by light travelling along all possible rays from a point source within a crystal in a given time. In isotropic crystals, the ray velocity surface is a sphere.

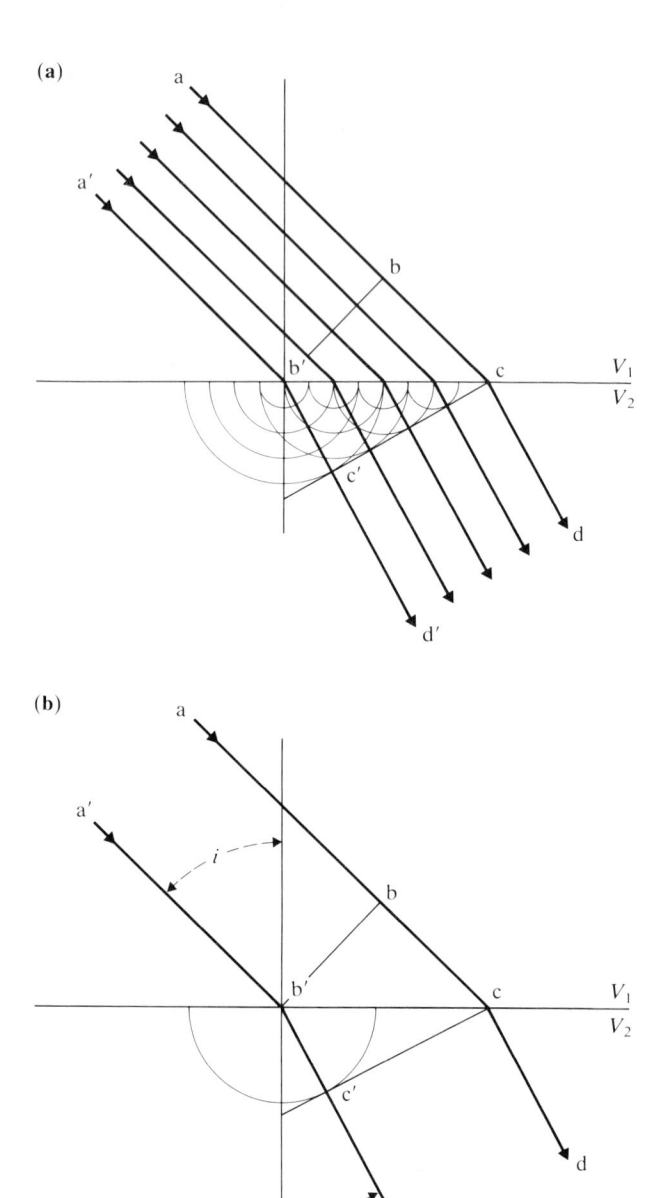

Figure 4.2 Refraction of light at a plane surface (a) Huygenian construction for several rays (b) simplified version of (a).

THE BIAXIAL INDICATRIX

Anisotropic crystals transmit light with different velocities in different directions, and the ray velocity surface of such a crystal is an ellipsoid, which may be of two principal geometric types, biaxial and uniaxial.

4.4 The biaxial indicatrix (triaxial ellipsoid)

Anisotropic crystals belonging to the orthorhombic, monoclinic and triclinic systems are *biaxial*, and are characterised by having three principal refractive indices and two optic axes. The relationship between the refractive indices can best be seen in a biaxial indicatrix, which is a triaxial ellipsoid possessing three planes of symmetry, with the three principal refractive indices equal to the three main semi-axes of the ellipsoid. These three refractive indices are given the symbols n_α, n_β and n_γ, and in *all* biaxial crystals n_γ is greater than n_β which is greater than n_α (i.e. $n_\gamma > n_\beta > n_\alpha$). In positive biaxial crystals n_β is closer in value to n_α, whereas in negative biaxial crystals n_β is closer in value to n_γ. Where n_β is exactly intermediate between n_α and n_γ it is impossible to determine whether the crystal is positive or negative (since 2V would be exactly 90°). Through any biaxial indicatrix two cross sections can be drawn which are true circles with a radius of n_β, because n_β is intermediate in size between n_α and n_γ. The position occupied in the indicatrix by these two circular sections depends on the relationship of n_β to n_α and n_γ, i.e. whether n_β is nearer n_α or n_γ in value (Figs 4.3 & 4).

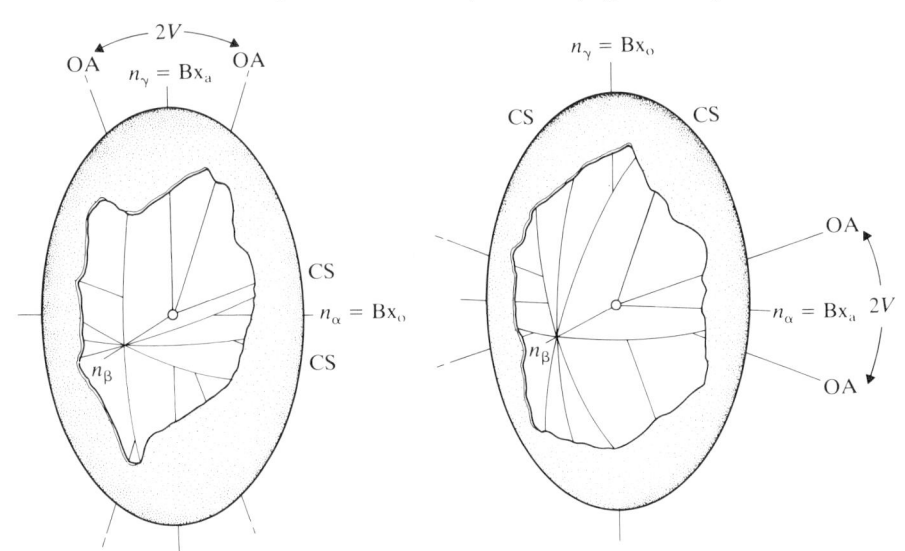

Figure 4.3 Positive biaxial indicatrices. OA, optic axis; CS, circular section; Bx_a, acute bisectrix; Bx_o, obtuse bisectrix.

Figure 4.4 Negative biaxial indicatrices. OA, optic axis; CS, circular section; Bx_a, acute bisectrix; Bx_o, obtuse bisectrix.

TRANSMITTED-LIGHT CRYSTALLOGRAPHY

Light travelling through a biaxial crystal in a direction which is perpendicular to a circular section will behave as if the crystal were isotropic. There are two circular sections of the ellipsoid, and therefore there must be two perpendiculars along which light will travel resulting in isotropic sections. These two perpendiculars to the circular sections are called the *optic axes* (OA) of a crystal, and this explains why the crystal is said to be biaxial. The optic axes lie in the plane of the ellipsoid containing the n_α and n_γ semi-axes. This plane is called the *optic axial plane* (OAP).

The optic axes may be arranged so that either n_α or n_γ bisects the smaller of the two angles between them. This smaller angle is called the *optic axial angle* or $2V$, and the semi-axis which bisects $2V$ is called the acute bisectrix or Bx_a. In Figure 4.3, n_γ is Bx_a since it bisects $2V$. The other semi-axis is called the obtuse bisectrix or Bx_o. In a positive crystal n_γ is Bx_a, whereas in a negative crystal n_α is Bx_a.

Light is polarised into two components on entering a biaxial crystal, and these components are shown (for light entering a crystal) along each of the three semi-axes n_α, n_β and n_γ in Figure 4.5. A ray velocity surface can be constructed which represents the distance these components will travel in a given time, and this is shown in Figures 4.6 and 4.7.

4.5 The uniaxial indicatrix

Anisotropic crystals belonging to the tetragonal, trigonal and hexagonal crystal systems are uniaxial. In a uniaxial indicatrix, which is also an ellipsoid, the two horizontal semi-axes (represented by n_α and n_β in the biaxial indicatrix) are equal (i.e. $n_\alpha = n_\beta$) and the ellipsoid has a circular cross section. This can be regarded as a limiting case of the biaxial indicatrix. Uniaxial crystals are therefore represented by an ellipsoid

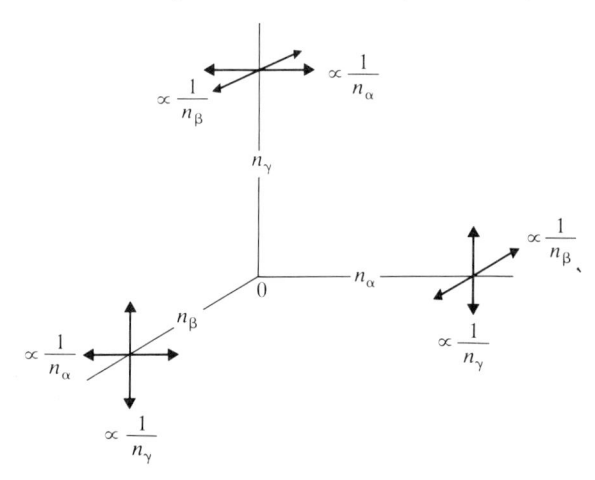

Figure 4.5 Polarisation in a biaxial crystal.

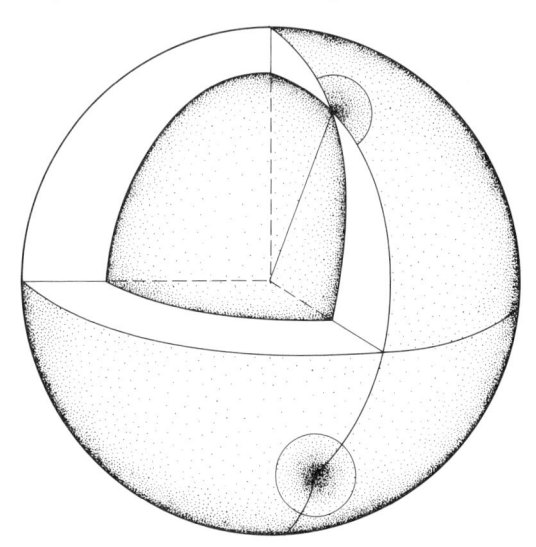

THE UNIAXIAL INDICATRIX

Figure 4.6 Ray velocity surfaces.

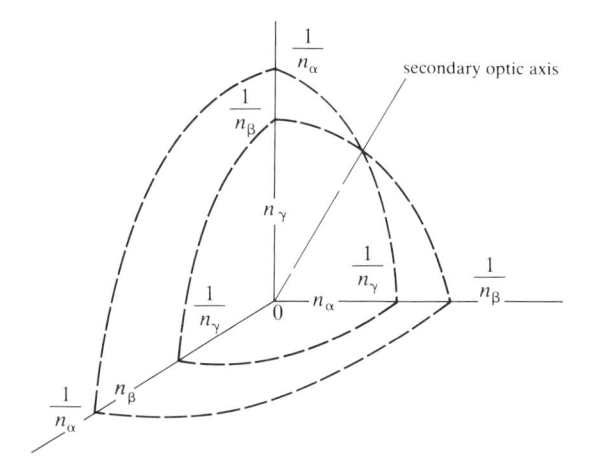

with a vertical semi-axis n_e, and a circular cross section of radius n_o. Thus n_e and n_o represent the two principal refractive indices of a uniaxial crystal. Two possibilities exist: either n_e is greater than n_o (termed positive), or n_e is less than n_o (termed negative) (Figs 4.8 & 9). Light travelling through a uniaxial crystal along the n_e direction (the vertical axis, perpendicular to the circular section) will behave as if the crystal were isotropic. In a uniaxial crystal n_e is always coincident with the c crystallographic axis, and therefore a crystal section cut at right angles to the c axis (a *basal section*) is isotropic.

In all other directions the crystal is anisotropic. Light entering the crystal along a horizontal radius n_o is polarised into two components

Figure 4.7 Ray velocity surfaces in three dimensions.

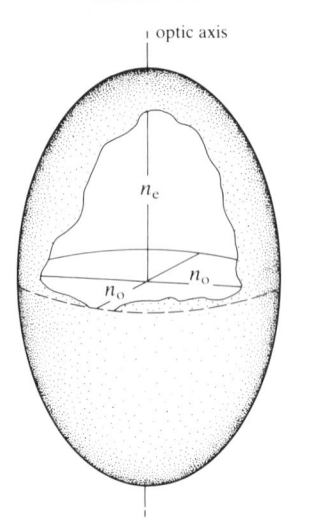

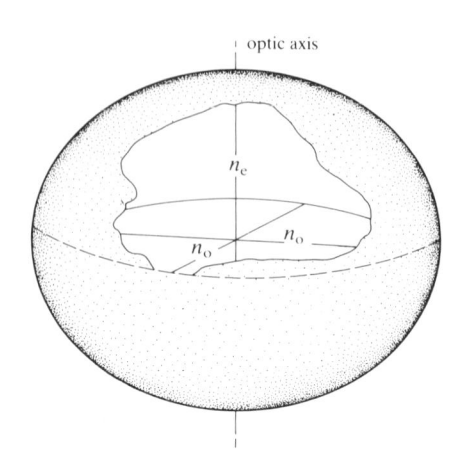

Figure 4.8 Positive uniaxial indicatrices. **Figure 4.9** Negative uniaxial indicatrices.

which vibrate in planes at right angles to each other, with velocities proportional to $1/n_o$ and $1/n_e$ (Fig. 4.10). Ray velocity surfaces can be drawn representing the distance that these components will travel in a given time, and these surfaces are shown for positive and negative uniaxial crystals in Figures 4.11 and 4.12.

4.6 Interference colours and Newton's Scale

Anisotropic crystal grains exhibit colours called interference colours when white light passes through them under crossed polars, provided that an optic axis is not parallel to the microscope axis, in which case the grain behaves as if it were isotropic. Constructive or destructive interference (i.e. brightness or darkness) of *monochromatic* light passing through the crystal fragment depends on the path difference between the two components, and the orientation of the planes of polarisation of the crystal in relation to the microscope polariser and analyser. If plane polarised light of a particular wavelength enters a crystal plate rotated from an extinction position, the monochromatic light is resolved into two components vibrating in mutually perpendicular planes (double refraction). The two components travel with different velocities through the crystal, and on emergence are not in phase. The path difference between them depends on the distance travelled through the crystal (i.e. the thickness of the crystal):

$$\text{path difference} = \frac{\Delta n t}{\lambda}$$

INTERFERENCE COLOURS AND NEWTON'S SCALE

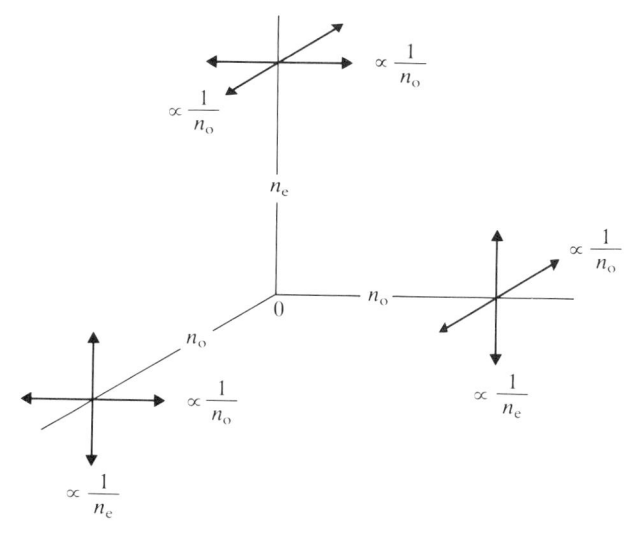

Figure 4.10 Polarisation in a uniaxial crystal.

where Δn is the birefringence of the crystal, that is, the difference between the maximum and minimum refractive indices, t is the thickness of the crystal in nanometres (1 μm = 1000 nm) and λ is also in nanometres. The path difference, as defined in the equation above, is expressed in fractions or whole wavelengths. The value Δnt is known as the retardation and is expressed in nanometres. The two components are combined into a resultant wave as the light passes through the analyser.

If the path difference is $m\lambda$, where m is a whole number, the waves combined by the upper analyser are $(m/2)\lambda$ out of phase (where m is an odd number). This is because the polariser and analyser are at 90° to each other. Such waves are similar in amplitude and in opposition

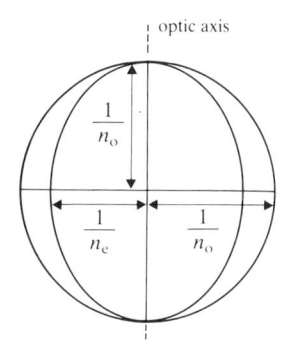

Figure 4.11 Principal sections of positive ray velocity surfaces.

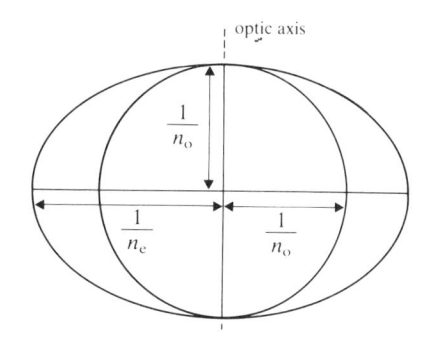

Figure 4.12 Principal sections of negative ray velocity surfaces.

Figure 4.13 Destructive inferference.

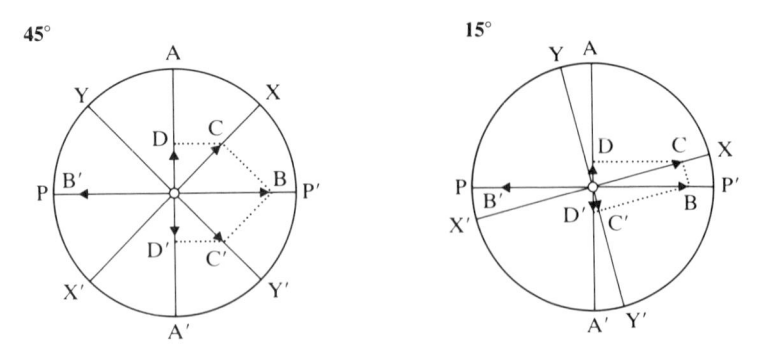

whatever the angular position of the crystal section, and the result is a wave of zero amplitude (destructive interference). In Figure 4.13, PP' is the polariser transmission plane, AA' is the analyser transmission plane, XX' and YY' are the two components into which light is resolved on passing through the crystal, OB is the amplitude of the wave leaving the polariser, OC and OC' are the amplitudes of the two components after passing through the crystal plate, and OD and OD' are the amplitudes of these components resolved in the analyser transmission plane. When the crystal is in the 45° position in Figure 4.13, OD and OD' are equal and opposite and yield a resultant wave of zero amplitude. In the 15° position, although components OC and OC' are dissimilar, OD and OD' are equal and opposite and a wave of zero amplitude again results.

If the path difference is not $m\lambda$, say $m_2\lambda$, where m is an odd number, then the components transmitted by the analyser are in phase and superimposed, so that a maximum resultant wave is produced with the crystal in the 45° position which has twice the amplitude of either of the interfering waves. The intensity of light of this resultant wave is four times as great as the intensity of the light of either wave because intensity is proportional to square of amplitude. This case is illustrated in Figure 4.14. In this figure the notation is as before. This time, however, the components reinforce in the analyser transmission plane. In the 45° position in Figure 4.14, OD' and OD are equal and coincident, and

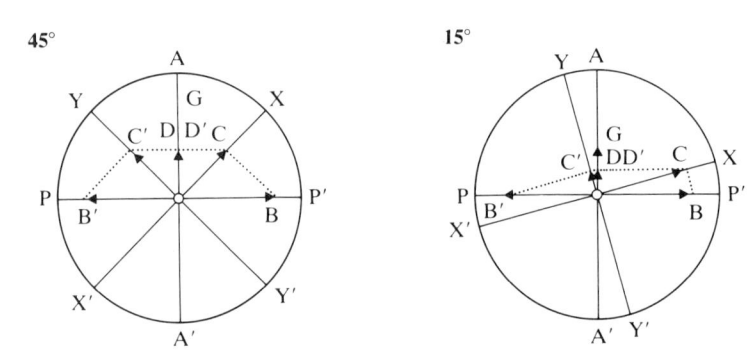

Figure 4.14 Constructive interference.

therefore analyser transmitted amplitude is 2OD. In the 15° position, although components OC and OC' are dissimilar, the components OD and OD' are still equal and coincident and a wave of amplitude 2OD again results. In Figure 4.14, the analyser transmitted amplitude is OG = OD + OD'.

In the 0° position XX' and YY' are coincident with PP' and AA' respectively. The components have no value (since D and D' would be coincident with 0) and therefore extinction would result. Thus from the above discussion resultant waves have a maximum amplitude (and maximum light intensity) in the 45° position, a smaller amplitude in the 15° position and zero amplitude in the parallel position; in this way a mineral will extinguish four times during a complete (360°) rotation of the microscope stage. Any path difference produced by a crystal fragment results in illumination, but the intensity decreases as the path difference approaches the wavelength.

The origin of interference colours can best be understood by considering the quartz wedge, remembering that varying the thickness of a crystal plate produces a variation in the path difference (or retardation). If a wedge cut parallel to the c axis of a crystal of quartz ($\Delta n = 0.009$) is inserted into the path of *monochromatic* sodium light passing through a microscope, then bright yellow bands are seen where the thickness of the wedge results in a path difference of $m\lambda/2$ with m an odd number, and dark bands are seen where m is an even number. The wavelength of sodium light is 580 nm, and therefore the bright yellow bands occur when $\Delta nt = 580 \times 1/2, 580 \times 3/2, 580 \times 5/2$ nm etc., and the dark bands occur when $\Delta nt = 580, 580 \times 2, 580 \times 3$ nm etc.

White light is composed of wavelengths ranging from 380 nm to 770 nm (violet to red). A quartz wedge inserted into the path of white light through a microscope produces a 'spectrum' of colours. Each different wavelength gives darkness and maximum intensity of colour for that wavelength at different positions along the wedge (Fig. 4.15). Overlapping of these various darknesses and maximum intensities combines to form a series of colours, known as Newton's Scale, which is shown in the colour chart (back cover). The colours are divided into different *orders*. The colours of the first order are black, grey, white, yellow and finally red. In the second order the colours are violet, blue, green, yellow, orange and red. Above this the colours become fainter, and the third order consists of indigo, green/blue, yellow, red and violet. Above the third order, mixing of wavelengths produces an easily identifiable pink colour.

In summary, the interference colour produced by an anisotropic mineral grain in a thin section depends on the retardation effect, which depends on the birefringence of the grain and its thickness.

Figure 4.15 Quartz wedge spectra.

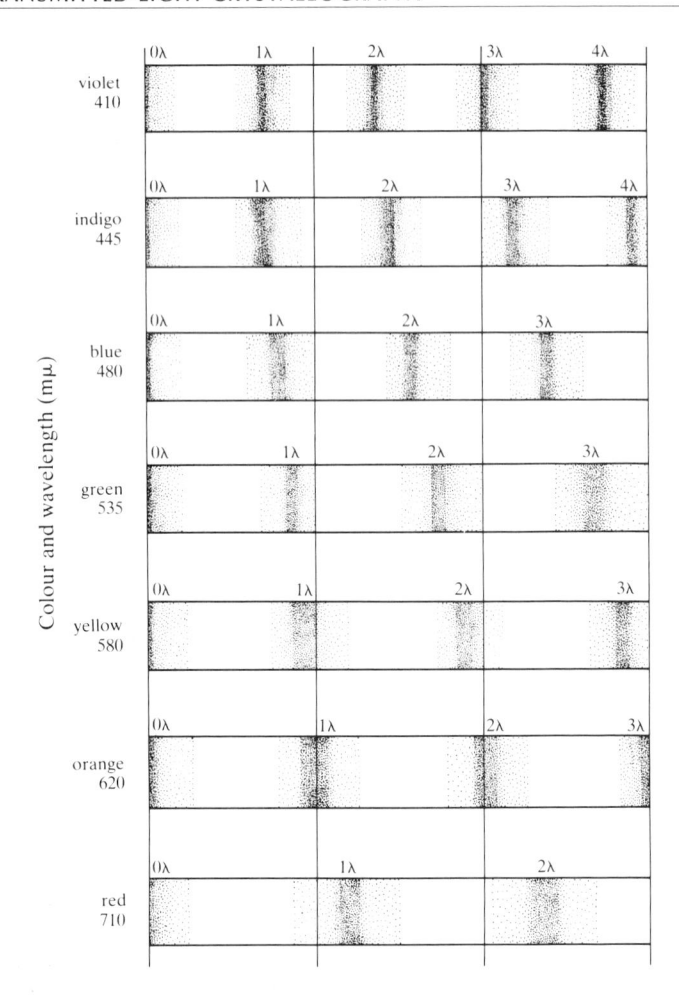

Colour and wavelength (mμ):
- violet 410
- indigo 445
- blue 480
- green 535
- yellow 580
- orange 620
- red 710

4.7 Fast and slow components, and order determination

4.7.1 Fast and slow components

In biaxial crystals the refractive indices are n_γ, n_β and n_α with $n_\gamma > n_\beta > n_\alpha$. Refractive index is inversely proportional to the velocity of light for a particular wavelength: therefore $1/n_\gamma < 1/n_\beta < 1/n_\alpha$. For example, along the vertical semi-axis of the biaxial indicatrix shown in Figure 4.5 the two components into which the light is polarised have velocities proportional to $1/n_\beta$ and $1/n_\alpha$. Because $1/n_\alpha$ is greater than $1/n_\beta$, the component with a velocity proportional to $1/n_\alpha$ will move further than the component with a velocity proportional to $1/n_\beta$ in a given time t. That is, one component ($\propto 1/n_\alpha$) is faster than the other ($\propto 1/n_\beta$), or one component is *fast* ($\propto 1/n_\alpha$) and the other *slow* ($\propto 1/n_\beta$).

In any anisotropic crystal, light is polarised into two components, one of which will be fast and the other slow. For example, in a positive uniaxial indicatrix, the vertical semi-axis n_e is greater in value than n_o, and therefore $1/n_e$ will be less than $1/n_o$. Light entering a positive uniaxial crystal at right angles to n_e will be split into two components, one of which has a velocity proportional to $1/n_e$ (slow) and the other with a velocity proportional to $1/n_o$ (fast). Since n_e is coincident with the c axis in uniaxial crystals, a section at right angles to the c axis is a *prism section*, and a prism section (of a positive uniaxial, mineral) is length slow. (A uniaxial negative mineral will be length fast on a prism section).

4.7.2 Quartz wedge and first order red accessory plate

These microscope accessories are length slow. The quartz wedge is cut with its length parallel to the prism zone. If the first order red plate (also called gypsum plate or sensitive tint plate) is examined under a microscope, with both analyser and polariser in position, the colour transmitted by the plate is red of the first order of Newton's Scale, corresponding to a retardation of 560 nm.

4.7.3 Determination of order of colour

To determine the maximum order of colour displayed by a mineral, a section showing maximum birefringence giving the highest order of interference colour is needed.

In biaxial crystals such a section has n_α and n_γ in its plane, and the section is at right angles to the semi-axis n_β. Light will be polarised into two components, one $\propto 1/n_\alpha$ (fast) and the other $\propto 1/n_\gamma$ (slow).

In a uniaxial crystal the section required is a prism section. To determine, in any mineral grain, which component is fast and which is slow, the procedure is as follows:

(1) Rotate the mineral being examined into extinction and make a sketch of the mineral in this position.
(2) The two components will then be parallel to the vibration directions of the polariser and analyser.
(3) Rotate the section through 45° and insert a first order red plate which will be parallel to one of the components. If this accessory plate is length slow, then *addition* (or increase in retardation) will occur if the component is also slow, i.e. the retardation of the plate will be added to the retardation of the mineral. For example a mineral having second order green interference colours will be changed to third order green. If the component lying parallel to the

plate is fast, then *subtraction* (or compensation) will occur and the retardation of the plate is subtracted from the retardation of the mineral. For example a mineral showing second order green will be changed to first order grey.

(4) The fast component identified from step 3 is brought parallel to the direction along which an accessory plate can be inserted. The Bertrand lens may be inserted. A length slow quartz wedge is inserted until a *black band* is obtained. In fact the colour may appear more like dark grey than black and represents the beginning of Newton's Scale. The black band marks the position where the retardation of the quartz wedge exactly compensates that of the mineral. The quartz wedge is then slowly withdrawn and the orders of colours counted until the original interference colour of the mineral section is restored. Alternatively the thin section may be removed and the order assessed by viewing the order of colour of the quartz wedge.

4.7.4 Abnormal or anomalous interference colours

These are obtained occasionally because a particular mineral is isotropic for a particular wavelength (or colour); this wavelength is removed and the complementary colour appears. Thus, for example, melilite and chlorite have yellow removed and appear dark blue.

4.8 Interference figures

4.8.1 Biaxial minerals

A mineral section with a vertical optic axial plane is first considered. If Bx_a is vertical, that is parallel to the microscopic axis, the situation is as follows:

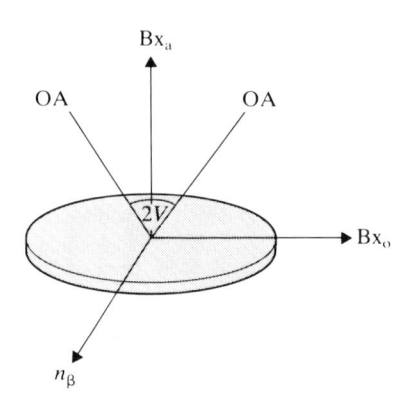

INTERFERENCE FIGURES

Note that n_β is always at right angles to the OAP. Light entering the crystal section is resolved into two components along each major semi-axis as follows:

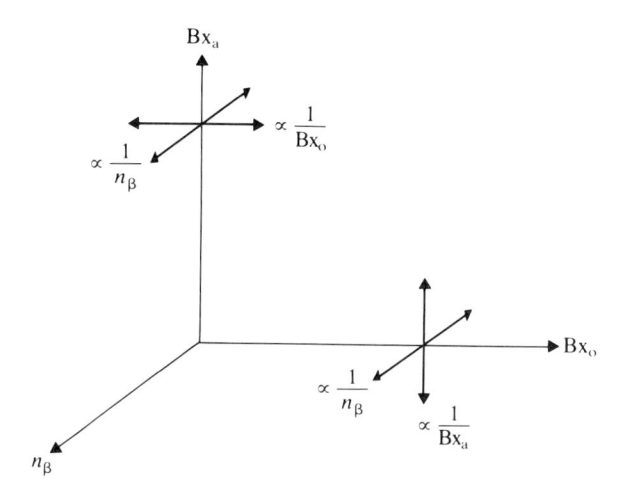

[diagram showing Bx_a and Bx_o axes with components $\propto \frac{1}{n_\beta}$ and $\propto \frac{1}{Bx_o}$, $\propto \frac{1}{Bx_a}$]

The crystal is isotropic along OA; along the optic axis light moves with a velocity proportional to $1/n_\beta$. Note that, if the mineral is +ve, Bx_a corresponds to n_γ and Bx_o to n_α, and *vice versa* if the mineral is −ve.

When a substage convergent lens is present in the optical train, as is usually the case, it has the effect of producing light rays which enter the crystal fragment at every angle from nearly horizontal to vertical:

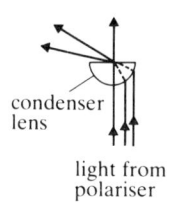

[diagram of condenser lens with light from polariser]

With this lens in position the plan view is as follows:

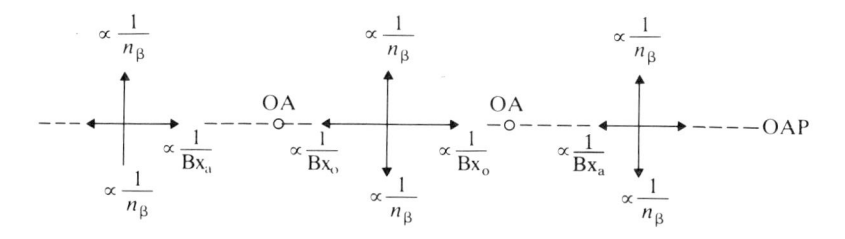

[diagram showing plan view with OA points and OAP direction, with components labeled $\propto \frac{1}{n_\beta}$, $\propto \frac{1}{Bx_a}$, $\propto \frac{1}{Bx_o}$]

In convergent light, the path difference produced by the double refraction of one ray is not equal to the path difference produced by another ray, and this results in a variation in interference effects across the field. In biaxial crystals *isogyres* (dark bands) and *isochromatic curves* (interference colours) appear, their behaviour depending upon the orientation of the crystal plate. Along the optic axes, no path difference occurs (0λ) and the optic axes appear as dark spots. Figure 4.16a shows isochromatic curves (black) for a path difference of 1λ in a section cut normal to Bx_a and Figure 4.16b shows the same curves in a thicker section. A set of curves mark loci of sets of points of emergence of components with a path difference of 1λ. The 'bright' curves between them are due to fractional path differences which occur within the centre of each 'bright' curve and correspond to a path difference of $m\lambda/2$, m odd.

These curves represent a slice through a three dimensional surface (known as Bertin's Surface) which is shown for a path difference of 1λ in Figure 4.17, with sections across it at various points. Other surfaces exist for different values of $m\lambda/2$, and a cross section through a set of these surfaces is shown in Figure 4.18; the cross section is perpendicular to Bx_a and both optic axes are seen.

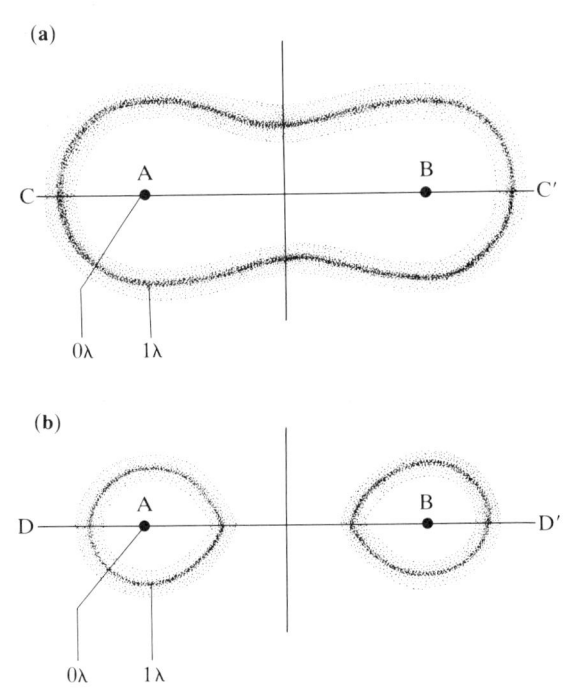

Figure 4.16 Isochromatic curves for path difference λ: (a) for section normal to Bx_a; (b) for thicker section.

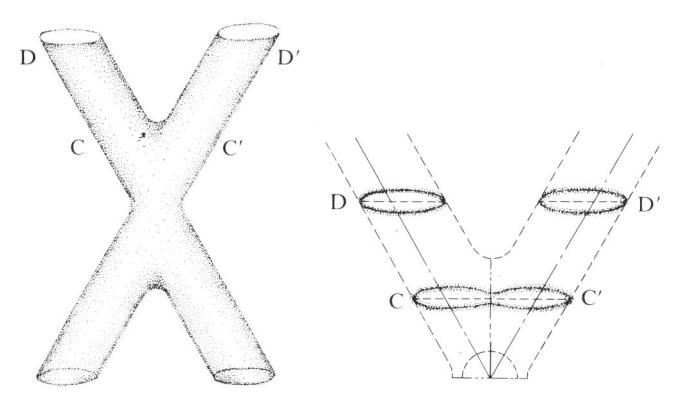

Figure 4.17 Bertin's Surface for path difference λ.

Interference figures from most minerals in thin section show only one or two isochromatic curves, corresponding to low first order interference colours.

Isogyres in biaxial crystals consist of dark curves determined by the loci of points of emergence of rays, the traces of whose planes of vibration are parallel to or nearly parallel to the planes of polarisation of the polariser and analyser. The black isogyres appear as hyperbolic curves. The stage is rotated until the isogyre is at 45° to the microscope crosswires and the curvature of the isogyre in this position gives an indication of the size of $2V$. If the curvature is great, $2V$ is small, whereas if the curvature is small (and the isogyres appear straight), $2V$ is large

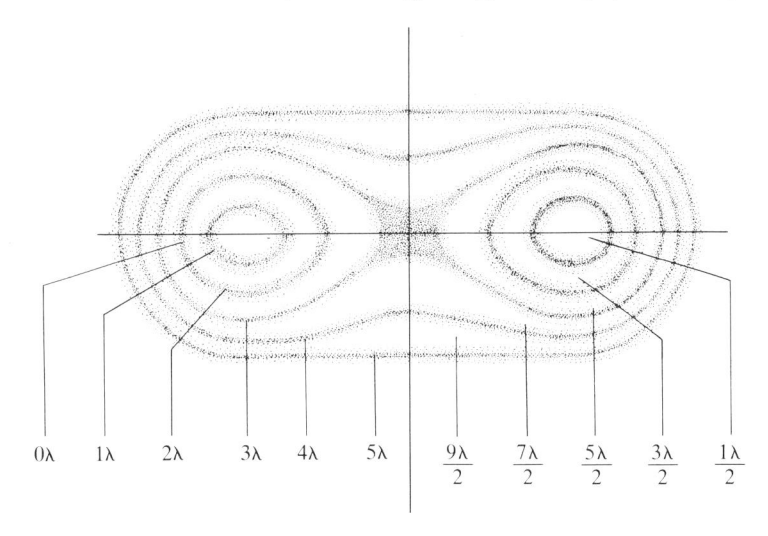

Figure 4.18 Cross-section through Bertin's Surface, showing constructive and destructive path differences.

(nearly 90°). The field of view of the microscope is such that for a biaxial Bx_a interference figure with a $2V$ of almost 40°, both optic axes can just be seen at the edge of the field of view. If the $2V$ is larger (greater than 40°), and especially when $2V$ nears 90°, a section cut normal to Bx_a is not suitable. This is because it is very difficult with a $2V$ greater than 70° to know whether the mineral section under investigation is cut perpendicular to Bx_a or Bx_o.

In this case a single optic axis interference figure must be obtained. This is done by finding an isotropic section which will have an optic axis vertical. The interference figure will show *one* optic axis with an isogyre passing through it.

4.8.2 Sign determination for biaxial minerals

The interference figure is rotated into the 45° position (Fig. 4.19). Either a single optic axis figure (for a large $2V$) or an acute bisectrix figure (for a small $2V$) is used.

The first order red plate (which we shall assume is length slow) is inserted along the OAP (see lower parts of Figs 4.20a, b, c & d). The sign of the component proportional to $1/Bx_a$ is obtained by observing the colour which appears on the *concave* side of the isogyre. *Blue* means that the mineral is positive (+ve). Addition has occurred since the length slow plate has been inserted along the slow component, which is proportional to $1/n_\gamma$ (i.e. $n_\gamma = Bx_a$). A yellow colour means that the mineral is negative (−ve) with $Bx_a = n_\alpha$. (Note that sign determinations are always made in the 45° position, with the first order red plate inserted along the OAP to find the relative velocity of the acute bisectrix component and hence obtain the sign.)

4.8.3 Flash figures

Sections cut parallel to the OAP (i.e. the OAP is in the plane of the mineral section) are perpendicular to n_β, and an interference figure called a *flash figure* is seen. It is possible to obtain the sign of a mineral from such a figure, but this is not recommended. A *flash figure* is obtained from a crystal fragment containing n_α and n_γ in the plane of the section.

Figure 4.19 Biaxial interference figure.

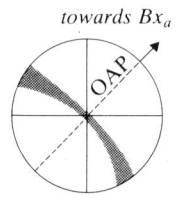

4.8.4 Uniaxial minerals

A uniaxial interference figure can be thought of as a special case of the biaxial figure where $2V = 0°$. When this occurs both optic axes become coincidental and the isochromatic curves appear as circles around the single optic axis.

The isogyres also coalesce and appear as a black cross, the arms of which are parallel to the polariser and analyser.

To determine the optic sign a centred cross should be used, but a slightly off-centred uniaxial interference figure may be rotated until the black cross is placed in the lower left hand quadrant of the field of view.

The length slow first order red plate is then inserted towards the cross, and the colour in the NE quadrant of the cross noted (Fig. 4.21). Blue signifies that the mineral is positive. 'Addition' of retardations has occurred since the length slow plate has been inserted parallel to the extraordinary ray direction. In uniaxial positive crystals $n_e > n_o$ and therefore $1/n_e$ is less than $1/n_o$. Yellow signifies that the mineral is negative and that 'subtraction' has occurred.

If a uniaxial mineral grain in the thin section is too small to allow an interference figure to be obtained, it should be noted that uniaxial positive minerals are always length slow parallel to the prism zone. A prismatic crystal should be observed under crossed polars with a low or moderate power objective. The mineral is rotated until the (length slow) first order red plate can be inserted along the prism zone of the crystal. If 'addition' occurs and the interference colour of the mineral is increased to a higher order of colour then the crystal is positive. If 'subtraction' occurs the crystal is negative.

In some cases the accessory plate may be length fast, and not length slow as described, and also the direction of insertion of the accessory plate may be NW to SE (and not NE to SW as described). The upper parts of Figure 4.20 explain sign determinations using a length slow *or* length fast first order red accessory plate which may be inserted either NE to SW *or* NW to SE into the field of view.

4.8.5 Isotropic minerals

Isotropic crystals do not show interference figures since they are isotropic to light in *all* directions. This fact enables us to distinguish isotropic sections of non-cubic minerals from those of cubic minerals.

4.9 Pleochroic scheme

4.9.1 Uniaxial minerals

The pleochroic scheme for a uniaxial coloured mineral (e.g. tourmaline) can be obtained by first finding a basal section to get the colour for n_o,

(a) Direction of insertion NE to SW (denoted ↙)

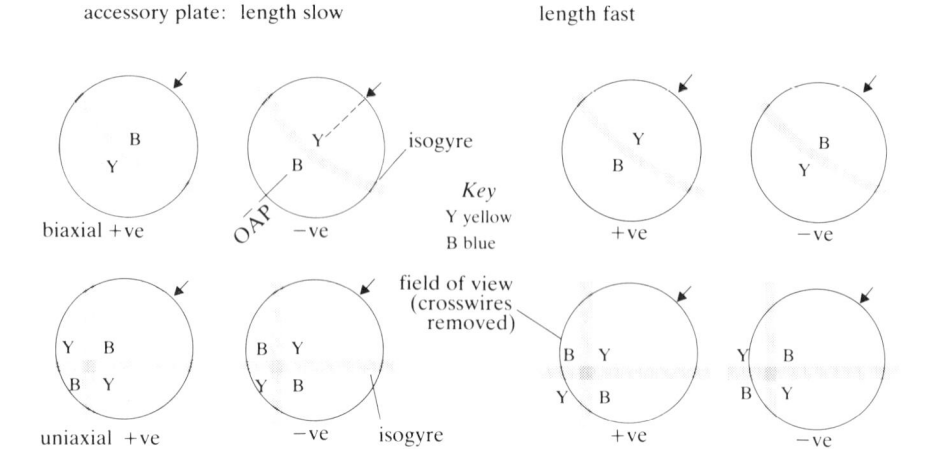

accessory plate: length slow — length fast

biaxial +ve | −ve | +ve | −ve

Key
Y yellow
B blue

field of view (crosswires removed)

uniaxial +ve | −ve | +ve | −ve

isogyre

(b) Direction of insertion NW to SE (denoted ↘)

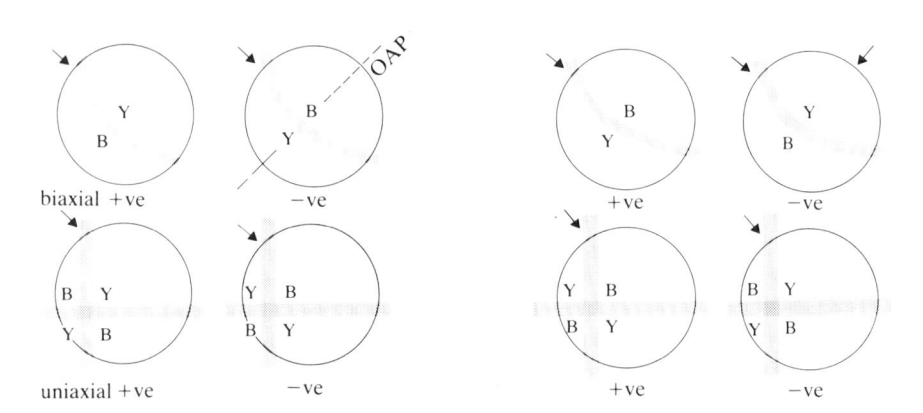

accessory plate: length slow — length fast

biaxial +ve | −ve | +ve | −ve

uniaxial +ve | −ve | +ve | −ve

Figure 4.20 Determination of optical sign of interference figures using either length slow or length fast accessory plates, and with their direction of insertion either NE–SW or NW–SE. In the sets of cartoons, numbers (a) and (d) are probably the most helpful. In (a) and (b), the uniaxial cross is placed in the lower left-hand corner of the field of view, and the biaxial single optic axis isogyre is rotated until it is concave towards the north-east. The cross and isogyre can, however, also be placed in the lower right-hand corner of the field of view with the isogyre concave towards the north-west. This is shown in (c) and (d).

(c) Direction of insertion NE to SW (denoted ↙)

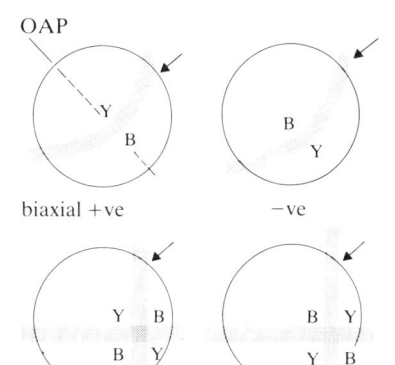

accessory plate: length slow   length fast

OAP

biaxial +ve   −ve   +ve   +ve

uniaxial +ve   −ve   −ve   −ve

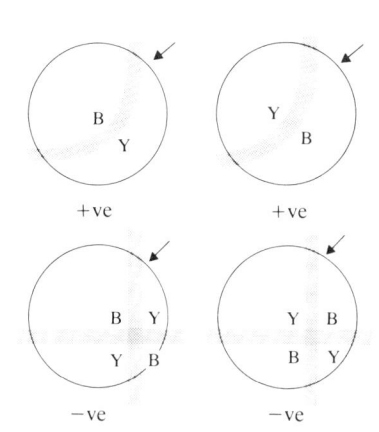

(d) Direction of insertion NW to SE (denoted ↘)

accessory plate: length slow   length fast

OAP

biaxial +ve   −ve   +ve   +ve

uniaxial +ve   −ve   −ve   −ve

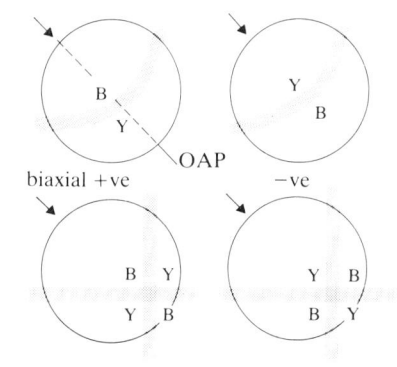

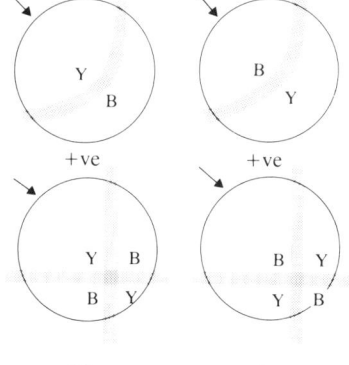

Figure 4.21 Uniaxial interference figure.

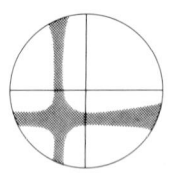

and then finding a prismatic section which is rotated so that the c axis (or optic axis) is lying east–west, to get the colour for n_e (see Fig. 4.5). The investigation is done using plane polarised light. The colour related to n_o is termed the o colour and that related to n_e termed the e colour.

4.9.2 Biaxial minerals

The pleochroic scheme for a biaxial crystal requires two differently oriented sections. An optic axis normal section which is isotropic under crossed polars will give the colour for n_β in plane polarised light. Alternatively a Bx_a figure can be used to find the orientation of n_β and the colour found by rotating n_β into the east–west position. Remember n_β is always at right angles to the OAP. Next a section is obtained showing maximum birefringence under crossed polars. Such a section should have both n_α and n_γ in the plane of section and will have a flash figure as its interference figure. The nature of each component has to be determined accurately. The fast component has a velocity proportional to $1/n_\alpha$ (and is called α), whereas the slow component has a velocity proportional to $1/n_\gamma$ (and is called γ). Identification is as follows:

(1) The section showing maximum birefringence is put into extinction, and the two components are now parallel to the polariser and analyser of the microscope.
(2) The section is rotated through 45° so that a length slow first order red plate can be inserted along one of the components. If addition of retardations occurs and the colour displayed by the mineral changes to a higher order, then the slow component of the mineral (i.e. proportional to $1/n_\gamma$) is parallel to the length slow direction of the plate. If subtraction occurs and the interference colour is reduced then the fast component (i.e. proportional to $1/n_\alpha$) of the mineral is in position.
(3) Each component, fast and slow, is rotated in turn into an east–west position and the colour noted in plane polarised light to get α and γ respectively.

4.10 Extinction angle

In anisotropic minerals the *extinction position* is always noted. This is the relationship of some physical property of the mineral—cleavage trace, face edge, twin plane – to the microscope crosswires when the mineral is in extinction.

EXTINCTION ANGLE

All uniaxial and orthorhombic biaxial minerals have *straight extinction* – that is, under crossed polars the mineral is in extinction when a prismatic or basal cleavage or prism edge is parallel to one of the crosswires. Other biaxial minerals possess *oblique extinction*, although in some minerals the angular displacement (between, for example, cleavage and crosswire) may be very small or zero, depending upon the orientation. The angular displacement is called the *extinction angle*, and is usually denoted γ (slow ray) or α (fast ray) to cleavage.

The mineral section is put into extinction and the character of the two components, which are parallel to the crosswires, noted by rotating each component into the 45° position and determining whether the component is fast or slow by using an accessory plate.

In many biaxial minerals a *maximum extinction angle* will be obtained from a section showing maximum birefringence. Such a section will have α and γ in the plane of the section, and the relationship of one of these components to a cleavage or other physical property can be determined. Note that the results of several readings on different grains are *not* averaged but that the *maximum* extinction angle is taken. A few biaxial minerals give a maximum extinction angle in a section which does not show maximum birefringence. In particular pigeonite, crossite, katophorite, arfvedsonite and kyanite show this, and the data for these minerals are given in Table 4.1.

Throughout the mineral descriptions in Chapter 2, large variations in the extinction angle may occur for a particular mineral. Such variations are due to changes in mineral chemistry, for example variations in $Mg:Fe^{2+}$ or $Ti:Fe^{3+}$ ratios in those minerals.

This chapter provides more detailed information on the passage of light through crystals than was given in Chapter 1. However, advanced texts such as Bloss (1971) will provide much greater detail on optical crystallography and the passage of light through crystals, and these are recommended to the reader.

Table 4.1 Extinction angle sections not coincident with maximum birefringence sections.

Mineral	Section for maximum extinction angle	Components in section	Extinction angle
pigeonite	‖ to (010)	γ–β	$\gamma\char`\^ cl = 37$–$44°$
crossite	‖ to (010)	α–β	$\beta\char`\^ cl = 3$–$21°$
katophorite	‖ to (010)	α–β	$\begin{cases} \beta\char`\^ cl = 20\text{–}54° \\ \alpha\char`\^ cl = 70\text{–}36° \end{cases}$
arfvedsonite	‖ to (010)	α–β	$\alpha\char`\^ cl = 0$–$30°$
kyanite	‖ to (100)	γ–β	$\begin{cases} \gamma\char`\^ \text{prismatic cl} = 30° \\ \beta\char`\^ \text{basal cl} = 30° \end{cases}$

5 Reflected-light theory

5.1 Introduction

The nature of polarised light is described in Section 4.1, which should be referred to if the reader is uncertain about what is meant by 'polarised' light.

In order to understand the optical properties of minerals in reflected light it is necessary to consider elliptically polarised light as well as linearly (or plane) polarised light. The concept of polarisation of light is discussed in detail in Galopin and Henry (1972), but the brief simplified and idealised account presented here should be adequate for beginners. The three categories of polarised monochromatic light are illustrated in Figure 5.1 and are named according to the nature of the cross section of the wave when viewed along the path of the ray. Vibration of a particle up and down to produce a wave confined to a plane is easy to visualise, but this is not true of vibration leading to ellipticity. Elliptically polarised light may be considered to consist of two linearly polarised components which are out of phase and vibrate at right angles. Elliptically polarised light can only be partially extinguished by rotating a polariser in its path, whereas linearly polarised light is completely extinguished when its vibration direction is normal to that of the polariser. Circularly polarised light is a special case of elliptically polarised light where the two components have the same amplitude and a path difference of one-quarter or three-quarters of a wavelength.

In reflected light microscopy we are dealing with normally incident linearly polarised white light, but the light reflected from the polished surface only remains lineary polarised in certain cases; all sections of cubic minerals and some sections of non-cubic minerals in certain orientations yield reflected linearly polarised light (see Fig. 5.1). On arriving at the surface of a polished section of an anisotropic ore mineral rotated from extinction, the linearly polarised white light can be considered to separate into two coherent components (see Section 5.3.3). On leaving the surface the two components recombine and the ratio of their amplitudes and their possible phase difference results generally in elliptically polarised light. The light reflected from ore minerals appears as 'white' light whose brightness and colour depend on the optical properties of the mineral (Sections 5.1.1, 5.2). This 'white' light consists of a mixture of coherent rays of all wavelengths of visible light, but each wavelength may differ in intensity and azimuth and nature of polarisation. We can only tell that the reflected light is rather complex by inserting and rotating the analyser and interpreting the resulting observations (see Section 5.3).

INTRODUCTION

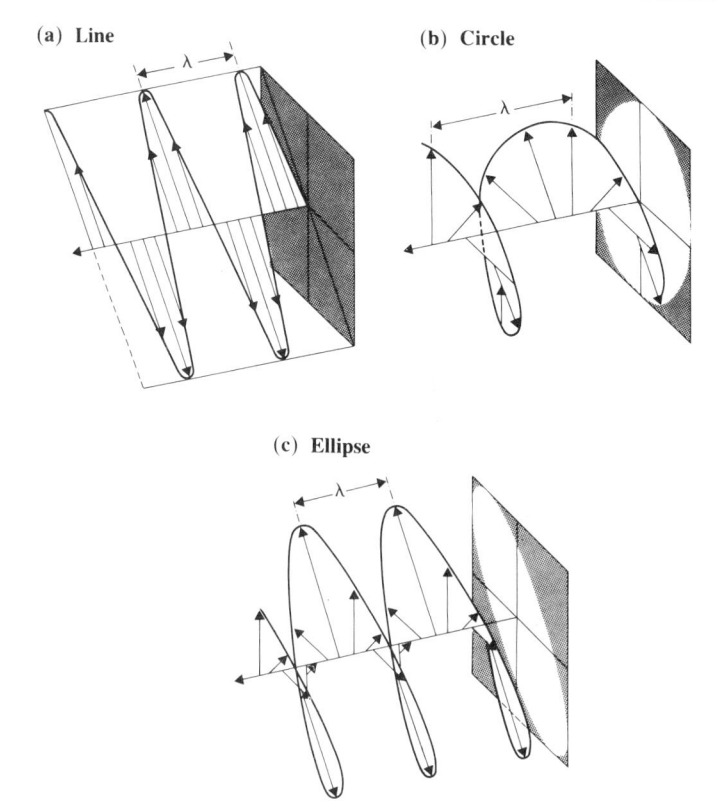

(a) Line **(b) Circle**

(c) Ellipse

Figure 5.1 Three categories of polarised monochromatic light.

5.1.1 Reflectance

The brightness of a mineral observed using reflected light microscopy depends of course on factors such as the intensity of the source lamp, but it also depends on the property known as reflectance. The reflectance of a polished section of a mineral is defined as the percentage of incident light that is reflected from the surface of the section. This reflected light travels back up through the objective of the microscope and eventually reaches the observer's eyes.

The reflectance of a mineral is not simply a single number; it depends on variables such as the crystallographic orientation of the section through the mineral and the immersion medium between the specimen and the objective. Reflectance is related to two fundamental properties, namely the optical constants termed the refractive index and the absorption coefficient. The relationship is expressed in the Fresnel equation:

$$R_\lambda \% = \frac{(n_\lambda - N_\lambda)^2 + k_\lambda^2}{(n_\lambda + N_\lambda)^2 + k_\lambda^2} \times \frac{100}{1}$$

where, for a wavelength value (λ), $R\%$ is the percentage reflectance, n is a refractive index of the mineral, k is an absorption coefficient of the mineral and N is the refractive index of the immersion medium.

The equation is strictly for reflection of linearly polarised light under normal incidence. It simplifies for observations in air where $N = 1$ for all wavelengths and for transparent minerals where $k = 0$.

The dispersion of the optical properties (i.e. their variation with wavelength) is much more important in understanding minerals in reflected light than in transmitted light.

The *refractive index (n)* and its variation with crystallographic orientation is dealt with in the theory of optical mineralogy for transmitted light studies (Section 4.2). However, it is worth noting that opaque minerals also have a refractive index.

The *absorption coefficient (k)* is a measure of opacity. As light of a given wavelength passes through matter it is progressively absorbed and the decrease in intensity is related to the absorption coefficient in the equation:

$$A = A_0 e^{-2\pi k d/\lambda_0}$$

where A_0 is the initial amplitude of a wave of wavelength λ_0, A is the amplitude after traversing a distance d in the crystal and e is the base of natural logarithms.

The intensity of a light wave is the square of the amplitude:

$$I = A^2$$

A mineral will appear opaque in thin section (0.03 mm thick) if its absorption coefficient is 0.01 or greater. The absorption coefficient varies with crystallographic orientation in the same way as the refractive index. Thus, for cubic minerals there is one refractive index (n) and one absorption coefficient (k); for uniaxial minerals the appropriate symbols are $n_o \neq n_e$ and $k_o \neq k_e$; and for lower symmetry minerals $n_\alpha < n_\beta < n_\gamma$ and $k_\alpha < k_\beta < k_\gamma$.

The relationship between the optical constants and reflectance and the variation with wavelength is shown using hexagonal pyrrhotite as an example in Figure 5.2. Remembering that the Fresnel equation holds at *each* wavelength, it can be seen how the *spectral reflectance curves* of pyrrhotite are related to the *dispersion curves of the optical constants*. It is because of the variation of reflectance with wavelength that pyrrhotite appears slightly coloured in polished section. More will be said on the colour of minerals in reflected light later.

Although an understanding of spectral reflectance curves is useful in the qualitative examination of minerals in polished section, a thorough treatment of theoretical aspects of reflected light and the measurement of reflectance and optical constants is unnecessary at this level, and the

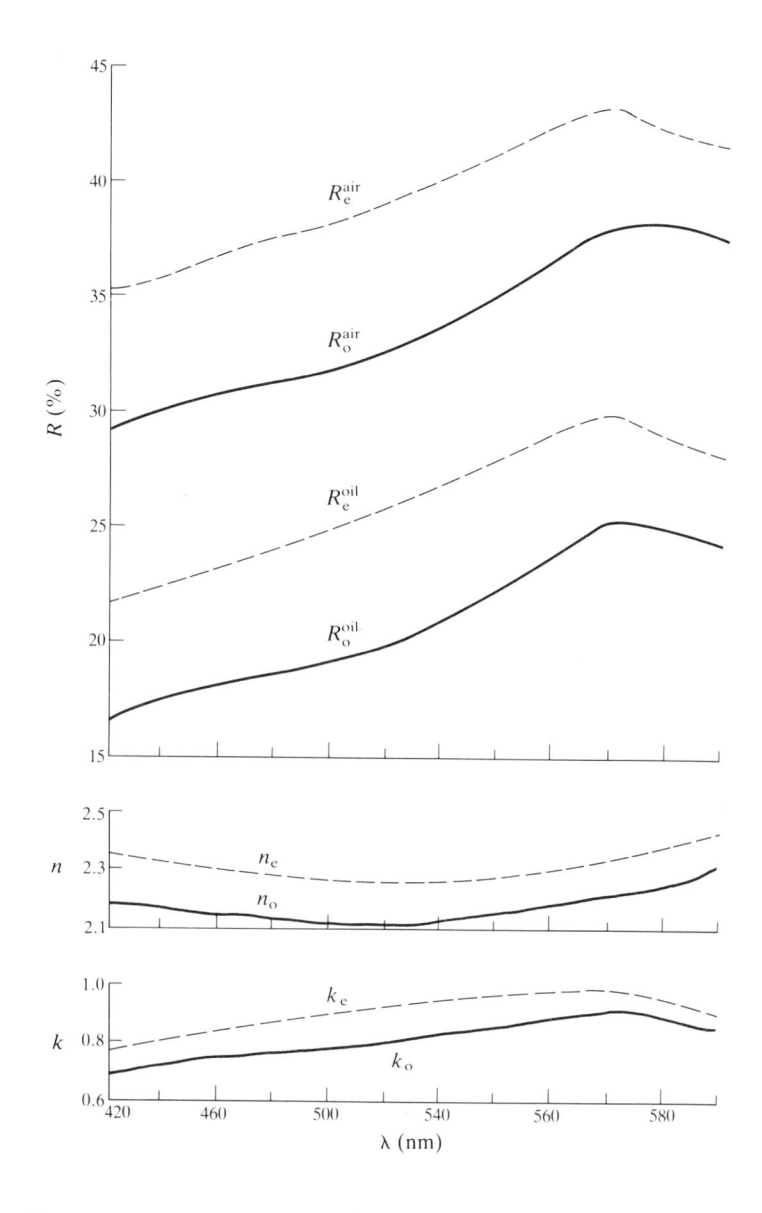

Figure 5.2 Variation with wavelength of $R\%$, n and k for hexagonal pyrrhotite (Cervelle 1979).

interested reader is referred to the textbook by Galopin and Henry (1972).

Returning to the Fresnel equation, it is worth noting that this equation explains why opaque minerals appear 'bright' in polished section. Although the reflectance of a transparent mineral increases with refractive index, a small increase in the absorption coefficient (i.e. opacity) leads to a marked increase in reflectance.

Examples of the relationship between refractive index, absorption coefficient and reflectance are shown for a range of minerals in Table 1.1. These examples emphasise the continuity in optical properties from transparent minerals, through weakly absorbing minerals, to truly opaque minerals.

5.1.2 Indicating surfaces of reflectance

As outlined above, the reflectance of minerals varies with crystallographic orientation. The directional nature of the reflectance can be described using an indicating surface which is analogous to but not identical with the refractive index indicatrix. The geometrical relationship between indicating surfaces and crystal symmetry is illustrated in Figure 5.3. The simplest surface is that for the cubic system; there is no variation in reflectance with orientation, so the indicating surface is a sphere. The surface for uniaxial minerals is a surface of rotation about the c axis; there is usually only a slight departure from a truly ellipsoidal surface. There is no theoretically correct surface for lower symmetry minerals because only certain crystallographic orientations reflect *linearly* polarised light.

5.1.3 Observing the effects of crystallographic orientation on reflectance

We are now in a position to understand reflection of light from aggregates of grains of a mineral as observed using plane polarised light. Cubic minerals have one reflectance value and one colour; there is no variation from grain to grain or within one grain on rotating the stage. Uniaxial minerals may vary in appearance from grain to grain; on rotation of the stage it should be possible at some position to make two grains of differing orientation appear identical in brightness and colour. Sections normal to the c axis of uniaxial minerals do not vary on rotation of the stage. Most grains of lower symmetry minerals will vary in reflectance and perhaps colour on rotating the stage.

Although it is easy to explain the behaviour of a mineral in terms of its crystal symmetry it is certainly *not* easy, and in any case usually unnecessary, to determine crystal symmetry from polished sections of minerals. See Figure 5.4, where the symmetry of crystals is illustrated.

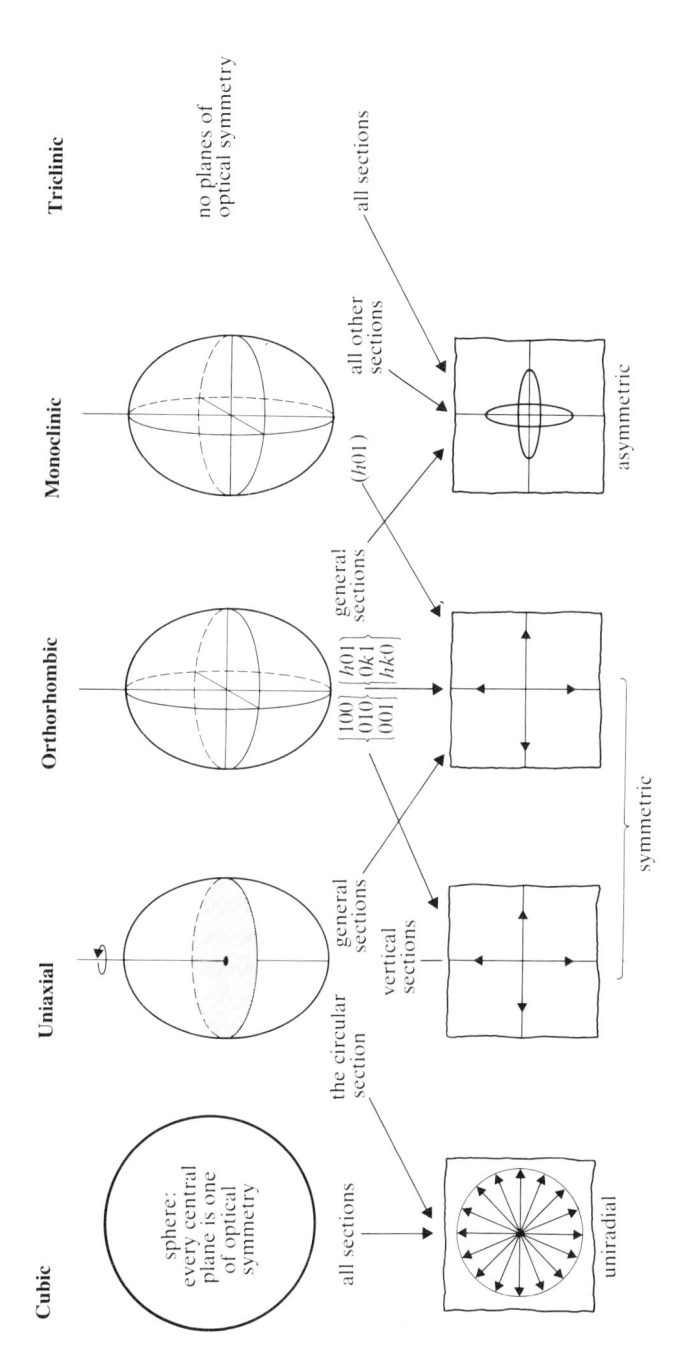

Figure 5.3 Indicating surfaces of reflectance. Considering sections in extinction positions, linearly polarised light is reflected from all sections of cubic and uniaxial minerals but only from certain sections (those normal to a symmetry plane) of orthorhombic and monoclinic minerals (after Galopin & Henry 1972).

Cubic
Isometric

octahedron 111   cube 100   dodecahedron 110

Uniaxial
Hexagonal

dipyramid $10\bar{1}1$ basal pinacoid 0001   rhombohedron $10\bar{1}1$
                          prism $10\bar{1}0$

Tetragonal

dipyramid 111   basal pinacoid 001   dipyramid 101
                 prism 110   prism 100

Low symmetry
Triclinic

pinacoids 111, $1\bar{1}1$, $1\bar{1}\bar{1}$, $11\bar{1}$ basal pinacoid 001   basal pinacoid 001
                                                               pinacoids $1\bar{1}0$, 110   pinacoids 010, 100

Monoclinic

basal pinacoid 001   basal pinacoid 001   pinacoid 101
front pinacoid 100   prism 110   pinacoid $10\bar{1}$
side pinacoid 010   side pinacoid 010

Orthorhombic

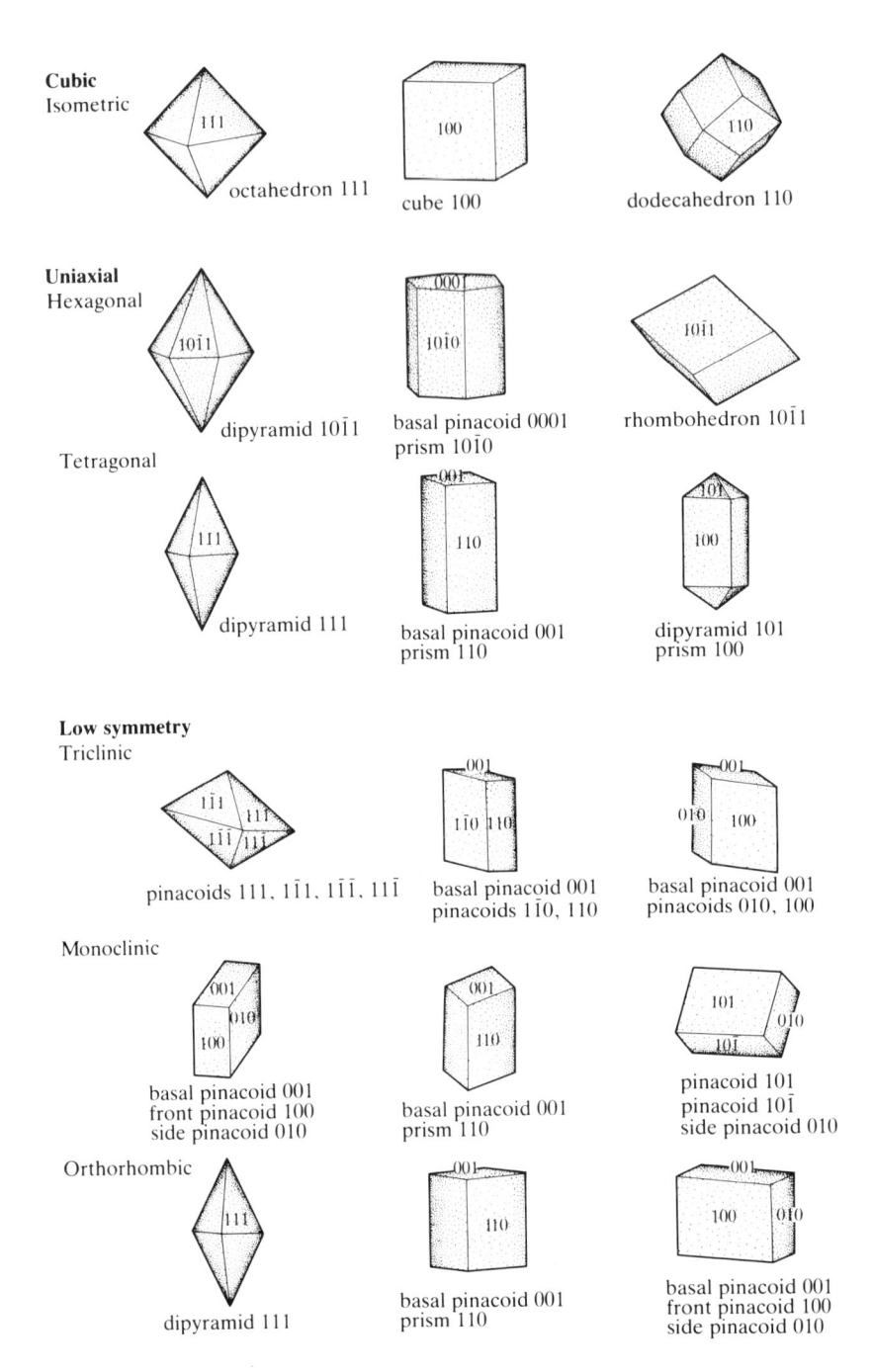

dipyramid 111   basal pinacoid 001   basal pinacoid 001
                 prism 110   front pinacoid 100
                                        side pinacoid 010

Figure 5.4 Crystal symmetry.

5.1.4 Identification of minerals using reflectance measurements

Accurate spectral reflectance curves are now available for all the common ore minerals. An identification scheme described by Bowie and Simpson (1980) is based on reflectance values at the four standard wavelengths 470, 546, 589 and 650 nm. Having measured several grains of the unknown mineral at the four standard wavelengths, the results are compared with the reference values on linear charts. If certain identification cannot be achieved immediately, then microhardness measurements or qualitative properties may be used to supplement the quantitative measurements. An advantage of their method is that although sophisticated research microscopes are required for accurate determination of spectral reflectance curves, relatively simple apparatus can be used to provide satisfactory reflectance values at the four standard wavelengths.

5.2 Colour of minerals in PPL

Recognition of the colour of minerals in polished section is useful in their identification, but unfortunately most minerals are only slightly coloured and the actual colours seen are easily changed. Colour change may be real, for example it may result from slight tarnishing; or it may be illusory, for example it may be caused by a varying background colour due to differing associated phases.

The application of quantitative colour theory to ore minerals has led to a better understanding of colour and its use in mineral identification. The colour perceived by the observer depends on the nature of the light source, the spectral reflectance curve of the mineral and how the observer interprets the spectral distribution of the light reaching his eye in terms of the mineral's surroundings. There is also the possibility of slight imperfection in the observer's colour vision. Obviously anyone who is severely colour blind is going to have great difficulties in using reflected light microscopy.

The quantitative colour system used is that of the Commission Internationale d'Eclairage 1931 (Judd 1952). If the standard data and their theoretical treatment are accepted, the only measurement required to obtain quantitative colour values of a polished section of a mineral is its spectral reflective curve. This curve represents the modification made by the polished surface of the mineral to the white colour of the light source. A surface with 100% reflectance at all visible wavelengths would obviously appear bright white (the colour of the source lamp). All minerals have reflectances much less than 100%, and since $R\%$ varies with wavelength this leads, but not in a simple way, to colour. Using the CIE (1931) colour diagram (described in detail in Section 5.2.1), minerals can be plotted and their colours compared quantitatively as well as

REFLECTED-LIGHT THEORY

qualitatively. Quantitative colour values of ore minerals are readily available in the IMA/COM DATA FILE (1977). They are presented as three numbers: visual brightness ($Y\%$), corresponding approximately to reflectance in white light; dominant wavelength (λ_d), which indicates the hue of the colour; and saturation ($P_e\%$), which indicates the strength of the colour. Thus bright white with a slight greenish tint would correspond to $Y\% = 50$, $\lambda_d = 585$, $P_e\% = 1$ and bright green to $Y\% = 45$, $\lambda_d = 585$, $P_e\% = 30$. Colour values vary for the type of source; only the A source (tungsten light) or C source (daylight) need be considered.

Cubic minerals have one reflectance curve and therefore one colour. A non-cubic mineral has a colour for each of its reflectance curves, and random sections are pleochroic but the pleochroism is usually very weak. Bireflection and pleochroism are closely related properties; the former is used when the only change seen is in brightness, whereas the latter is used if a change in colour, implying a change in dominant wavelength or saturation, is seen. Simple colour terminology, e.g. bluish white (not pale lavender blue!), should be used in mineral description.

It is important to emphasise that quantitative colour values can be used as an aid to mineral identification without the need for the observer to undertake spectral reflectance measurements. The use of quantitative colour values will soon be appreciated if the exercise in Section 5.2.2 is studied. An ore mineral identification scheme (NISOM1–81), based on quantitative colour measurements using a microcomputer interfaced to a reflected light microscope, has been developed and described by Atkin and Harvey (1982).

5.2.1 CIE (1931) colour diagram

All colours visible to the human eye under certain conditions plot in the colour diagram of Figure 5.5 within the field enclosed by the spectral locus (380 → 770 nm) and the 'purple line'. This area is two dimensional in terms of *colour*, but brightness can be plotted as a vertical axis and gives a three dimensional mountain-like body with 100% brightness (pure white) at point C (the colour of the source light) and 0% brightness around the perimeter. The colour of ore minerals plot within this mountain but they tend to plot in a zone from bluish through white to yellowish; there are few green minerals. As most minerals are only slightly coloured, they plot close to point C. Covellite (basal section) is plotted as an example; it is the 'deepest' blue mineral. Its approximate quantitative colour values (for C illuminant) are:

Covellite (R_o):   chromaticity co-ordinates $x = 0.224$ $y = 0.226$
dominant wavelength $= 475$ nm
% purity $= 42\%$
$Y\%$ $= 6.8\%$

COLOUR OF MINERALS IN PPL

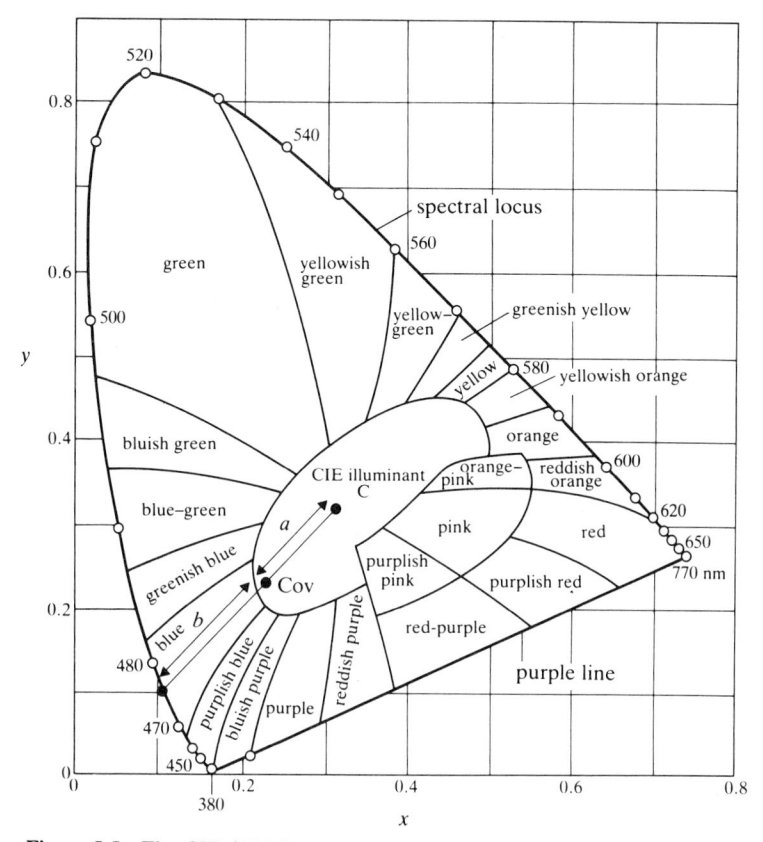

Figure 5.5 The CIE (1931) colour diagram, colour areas (Judd 1952).

Note that the dominant wavelength is given by a projection of a line from C through Cov to the spectral locus, and the % purity is given by the closeness of Cov to the spectral locus, i.e. $a/(a + b) \times 100$.

5.2.2 Exercise on quantitative colour values

Chromaticity co-ordinates and the visual brightness ($Y\%$) of an unknown mineral (B), sphalerite and the basal section of covellite are given on the CIE colour diagram Figure 5.6.

Plot mineral B on the diagram and explain, using quantitative colour values, how this mineral would appear in polished section. (Answer given at end of chapter.)

REFLECTED-LIGHT THEORY

Figure 5.6 Exercise on use of quantitative colour values: CIE colour diagram for A source.

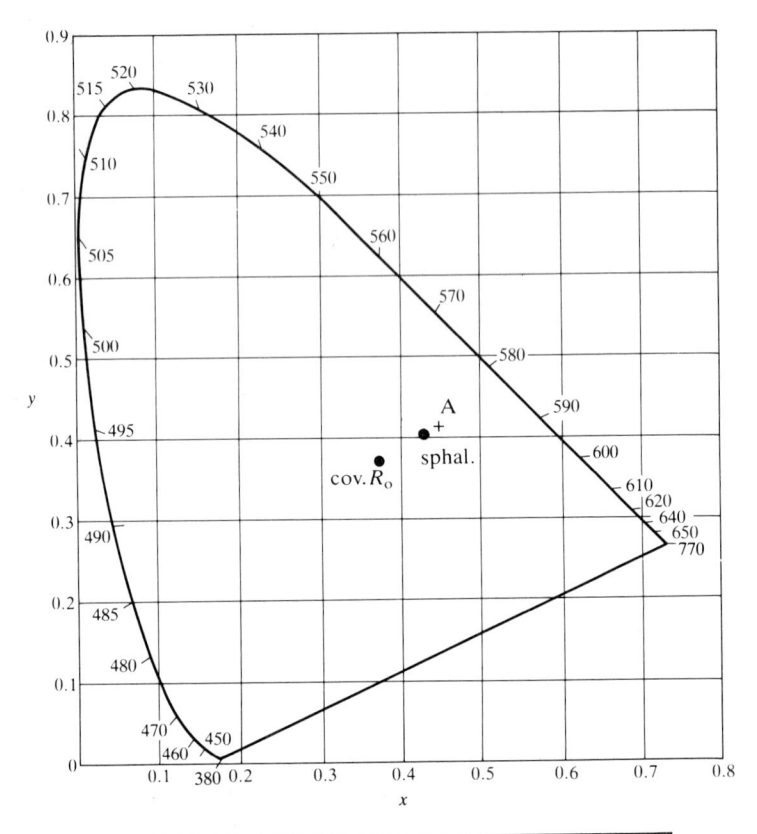

	Chromaticity co-ordinates		Y%
	x	y	
cov. R_o	0.370	0.370	7.0
sphal.	0.440	0.405	17.0
mineral B	0.400	0.385	20.0

5.3 Isotropic and anisotropic sections

5.3.1 *Isotropic sections*

Isotropic sections appear dark, ideally black, using crossed polars and they should not change in brightness on rotation of the stage. They will appear brighter and perhaps coloured if the analyser is slightly rotated, but again there should be no change in the appearance of the section as the stage is rotated. If all grains, i.e. small sections in different crystallographic orientation, of a mineral appear isotropic then the mineral is probably cubic. The mineral could however be non-cubic but with a very weak anisotropy. Basal sections of uniaxial minerals are isotropic.

5.3.2 Anisotropic sections

Anisotropic sections show colours, known as polarisation or anisotropic rotation colours, using crossed polars. The colour effects are usually weak, e.g. dark reddish browns or greys with a bluish tint. Anisotropy is best detected by using slightly uncrossed polars, but it must be remembered that this may change the polarisation colours. Some of the grains of a mineral will have a stronger anisotropy than others and some may be isotropic. Minerals exhibiting anisotropy are usually non-cubic, but cubic minerals may be distinctly anisotropic (e.g. pyrite).

Using exactly crossed polars, general sections of uniaxial minerals have four extinction positions at 90° and identical colours in each 45° quadrant. Even very slight misalignment of the polarising filters may change the colours, and for this reason the colours seen must be used with caution in mineral identification. Lower symmetry minerals also show polarisation colours but they need not have distinct extinction positions nor show the same colours in each 45° quadrant.

5.3.3 Polarisation colours

Polarisation colours differ in origin from interference colours seen in thin sections. Their origin can be explained with the help of Figure 5.7, which illustrates reflection from a uniaxial *transparent* mineral, such as calcite, in the 45° orientation. Incident linearly polarised monochromatic light, vibrating E–W, is resolved into two components, the two vibration directions (corresponding to extinction positions) on the surface of the section. On reflection, recombination of the components results in reflected linearly polarised light vibrating in a direction closer to the principal vibration direction of higher reflectance. The reflected light is now no longer vibrating normal to the analyser and some of the light will be able to pass through the analyser. Obviously the greater the difference between R_{max} and R_{min} the greater the angle of rotation, and this will result in more light passing through the analyser. As the angle of rotation may be dispersed, i.e. vary with wavelength, because the reflectance values of the principal vibration directions are dispersed, the amount of light of each wavelength passing through the analyser will vary, giving coloured light. The colours are usually weak because most of the light is cut out by the analyser.

Further complications arise in considering 'opaque' (absorbing) uniaxial minerals. Because of the different absorption coefficients (k) of the two principal vibration directions, the reflected light is no longer linearly polarised but elliptically polarised. The ellipticity results from

REFLECTED-LIGHT THEORY

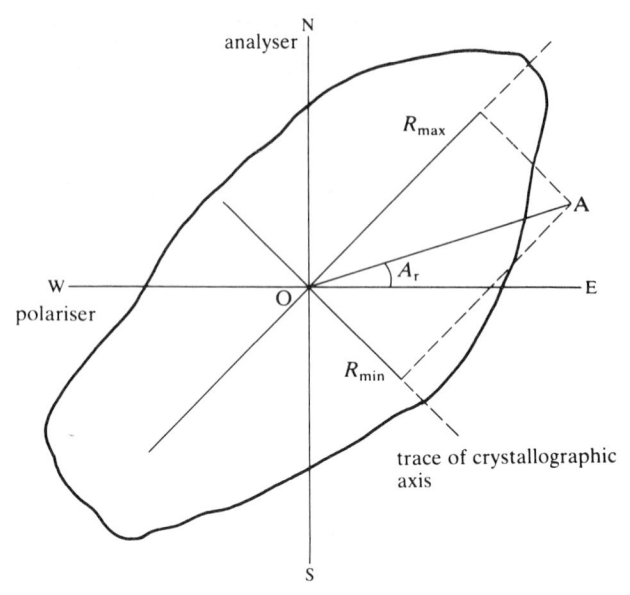

Figure 5.7 This figure illustrates the geometry of reflection of normal incident linearly polarised monochromatic light from a bireflecting (hkl) grain of a uniaxial transparent mineral turned to 45° from extinction. The incident light vibrating in the plane of the polariser (E–W) is resolved, on the polished surface, along the two principal axes R_{max} and R_{min} corresponding to n_{max} and n_{min}. This results in rotation of the plane of polarisation (i.e. the azimuth of vibration) through the angle A_r so that the reflected light is linearly polarised but vibrating parallel to OA.

the combination of two components of different magnitude and different phase. It is the difference in absorption which 'slows down' one component relative to the other and gives a phase difference. Some light will now pass through the analyser because of ellipticity as well as rotation. Dispersion of the degree of ellipticity contributes to colour effects.

So far only uniaxial minerals have been considered. The theory of reflection as outlined above only applies to lower symmetry minerals for sections normal to a symmetry plane; only these sections will contain two principal vibration directions that reflect linearly polarised light. General sections through low symmetry minerals reflect elliptically polarised light even from the section's principal vibration directions. Because of the possible differing crystallographic orientation of the three refractive indices and the three absorption coefficients of low symmetry absorbing minerals, and the dispersion of their orientation, the concept of optic axes and isotropic sections of biaxial transparent minerals does not have a simple analogy in the case of absorbing minerals (see Galopin & Henry 1972, p. 88).

ISOTROPIC AND ANISOTROPIC SECTIONS

5.3.4 Exercise on rotation after reflection

This exercise demonstrates the rotation of polarised light on reflection from an anisotropic grain, e.g. ilmenite, hematite (Fig. 5.8).

(1) Select a single optically homogeneous grain which exhibits distinct bireflectance and distinct anisotropy. Sketch the grain, positioned in one of the four extinction positions (found using exactly crossed polars), and indicate the reflectance values R_{max} and R_{min} of the

grain in brightest position (PPL) and extinction (x-polars). Reflected-light vibrates E-W.

grain in 45° orientation. Polarisation colour seen in x-polars.

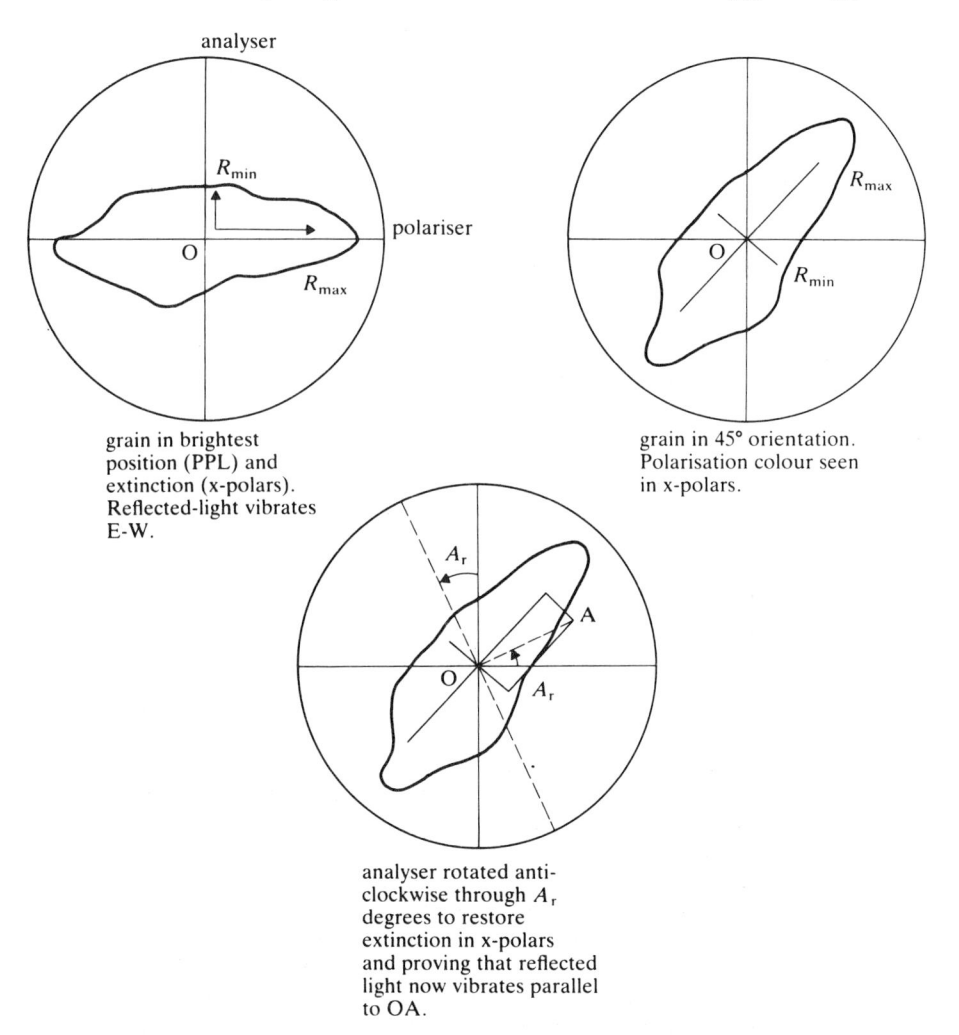

analyser rotated anti-clockwise through A_r degrees to restore extinction in x-polars and proving that reflected light now vibrates parallel to OA.

Figure 5.8 Exercise to demonstrate the rotation of polarised light on reflection from an anisotropic grain, e.g. ilmenite, hematite.

two vibration directions. Since the grain is distinctly bireflecting it should be possible to determine whether R_{max} lies E–W (grain at its brightest in PPL) or N–S (grain at its darkest in PPL).
(2) Rotate the grain exactly 45° from extinction so that R_{max} is directed NE–SW. Sketch the grain, again showing R_{max} and R_{min}, using a longer line for R_{max} to signify the greater percentage of light reflected.
(3) On your sketch, complete the rectangle to show the approximate vibration direction of the reflected light (OA in the figure).
(4) To prove that the light is in fact vibrating in this direction, push in the analyser and slowly rotate it a few degrees counter-clockwise (or rotate the polariser clockwise, 0° → 90°) until a position of darkness is obtained. This rotation causes the vibration direction of the analyser to become normal to the vibration direction of OA, so resulting in extinction. The 'apparent angle of rotation' A_r cannot be measured with sufficient accuracy to be of much use in identification using most student microscopes.
(5) Minerals showing strong pleochroism in PPL or vivid polarisation colours (e.g. covellite) display dispersion of the angle of rotation; on rotation of the analyser colours are obtained rather than a simple position of darkness. This display of colours is explained by extinction of some wavelengths of light at a given angle of rotation while others are transmitted to varying degrees. Slight movement of the analyser changes the distribution of extinguished and transmitted wavelengths. In a simple example, blue colours result from extinction of red light and vice versa.
(6) The origin of polarisation colours, as seen using exactly crossed polars, is explained in Section 5.3.3.

5.3.5 *Detailed observation of anisotropy*

Using exactly crossed polars, the 'strength' of anisotropy may be estimated from the amount of light reaching the eye with the section in the 45° orientation. If a mineral is strongly anisotropic then the anisotropy will be immediately evident if a group of grains of the mineral are examined and the stage rotated. The grain showing the strongest anisotropy can then be studied further to obtain additional information. It is important to ensure that the group of grains does represent one mineral!

The actual 'tints' seen with the polars exactly crossed should be noted. The vividness of the colours, i.e. how colourful they are, is an indication of the dispersion of the rotation angle and degree of ellipticity. Two examples may help in explanation: a bright grey colour represents strong anisotropy but small dispersion; a dark blue colour represents weak anisotropy but strong dispersion. 'Distinct' is a useful term to use because it indicates how easy it is to see the anisotropy.

ISOTROPIC AND ANISOTROPIC SECTIONS

Using slightly uncrossed polars the polarisation colours obtained will usually be sufficiently characteristic of the mineral to be useful in identification. A mineral showing four good extinction positions at 90° and the same tint 45° either side of an extinction position is probably uniaxial. If most sections show poor extinction and colours cannot be balanced about the 'best' extinction position the mineral is probably of lower symmetry.

The eye is best trained in the study of anisotropy by examining polished minerals of varying anisotropy and comparing observations with those given in standard tables.

Answer to problem in Section 5.2.2

Plot the mineral B on to the diagram using its chromaticity co-ordinates x and y. Draw a straight line from A through B to the spectral locus. All three minerals should lie on this line, and they have a dominant wavelength of 486 ± 4 nm. This means that B is bluish in colour and the hue (shade) of blue is exactly the same as covellite. The distance of a mineral from A towards the spectral locus indicates the purity (saturation or depth) of the colour. As sphalerite is essentially colourless and covellite is distinctly blue, we can say that B will be slightly bluish. The $Y\%$ values (brightness) approximate to $R\%$ for white light. Covellite is dark, sphalerite grey and mineral B is slightly brighter than sphalerite.

In summary, mineral B will appear in polished section as a slightly bluish light grey mineral; it will be slightly brighter than sphalerite and the blue colour will be of the same hue as the covellite basal section.

Appendix A.1 Refractive indices of biaxial minerals

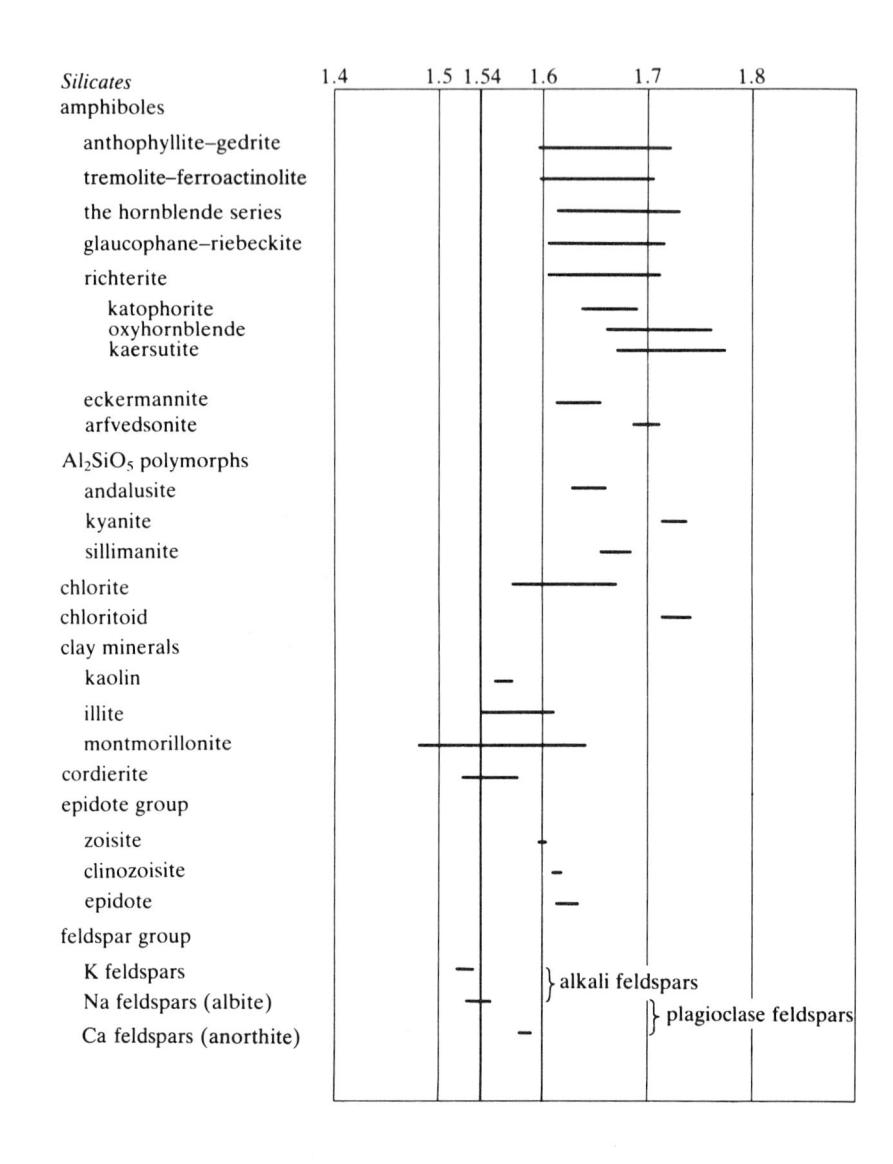

Mineral	1.4	1.5	1.54	1.6	1.7	1.8

Silicates
amphiboles
  anthophyllite–gedrite
  tremolite–ferroactinolite
  the hornblende series
  glaucophane–riebeckite
  richterite
    katophorite
    oxyhornblende
    kaersutite
  eckermannite
  arfvedsonite
Al_2SiO_5 polymorphs
  andalusite
  kyanite
  sillimanite
chlorite
chloritoid
clay minerals
  kaolin
  illite
  montmorillonite
cordierite
epidote group
  zoisite
  clinozoisite
  epidote
feldspar group
  K feldspars } alkali feldspars
  Na feldspars (albite)
  Ca feldspars (anorthite) } plagioclase feldspars

n	1.4	1.5	1.54	1.6	1.7	1.8	1.9

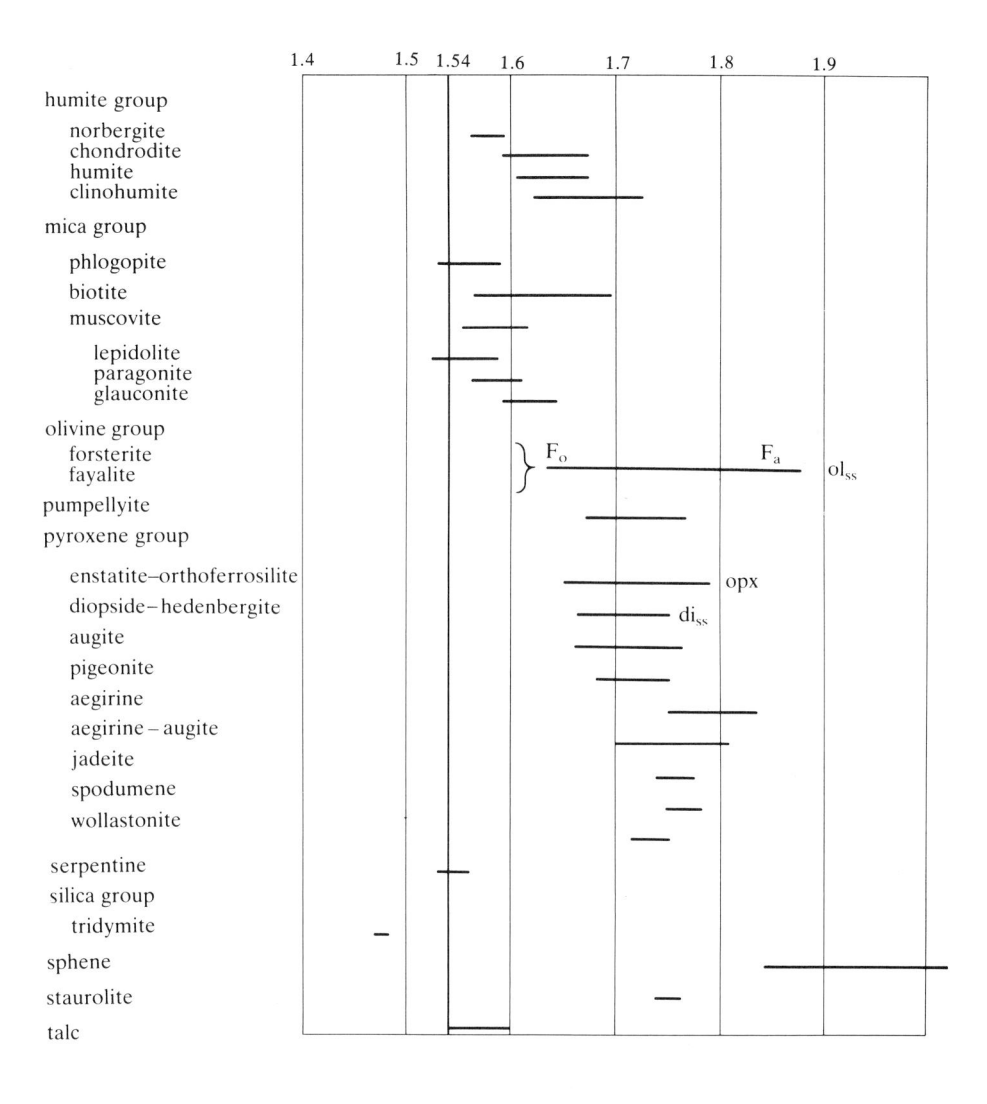

humite group
- norbergite
- chondrodite
- humite
- clinohumite

mica group
- phlogopite
- biotite
- muscovite
- lepidolite
- paragonite
- glauconite

olivine group
- forsterite F_o — F_a — ol_{ss}
- fayalite

pumpellyite

pyroxene group
- enstatite–orthoferrosilite — opx
- diopside–hedenbergite — di_{ss}
- augite
- pigeonite
- aegirine
- aegirine–augite
- jadeite
- spodumene
- wollastonite

serpentine

silica group
- tridymite

sphene

staurolite

talc

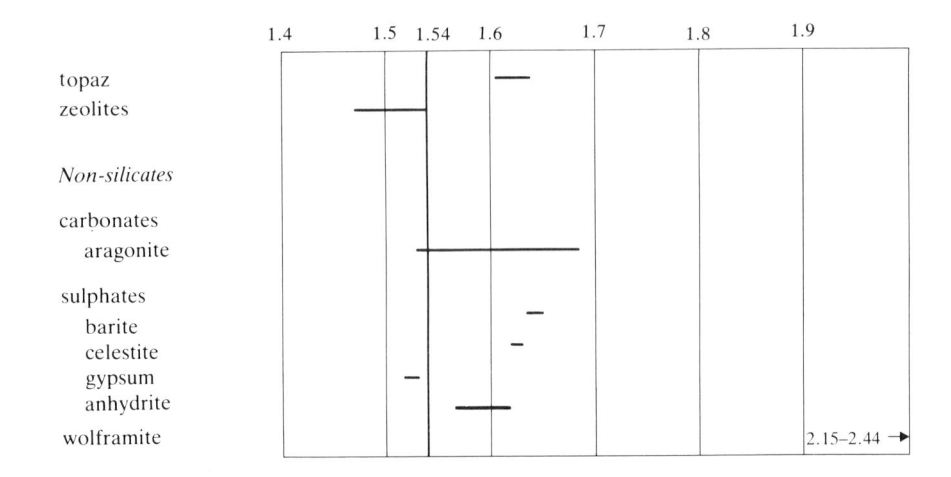

	1.4	1.5	1.54	1.6	1.7	1.8	1.9

topaz
zeolites

Non-silicates

carbonates
  aragonite

sulphates
  barite
  celestite
  gypsum
  anhydrite
wolframite — 2.15–2.44 →

Appendix A.2 Refractive indices of positive uniaxial minerals

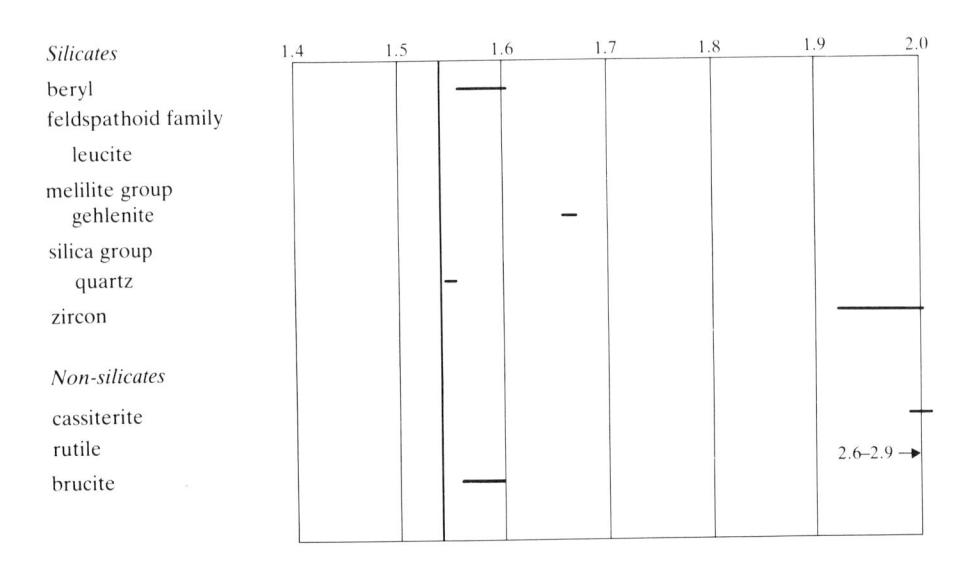

Silicates

beryl
feldspathoid family
  leucite
melilite group
  gehlenite
silica group
  quartz
zircon

Non-silicates

cassiterite
rutile — 2.6–2.9 →
brucite

Appendix A.3 Refractive indices of negative uniaxial minerals

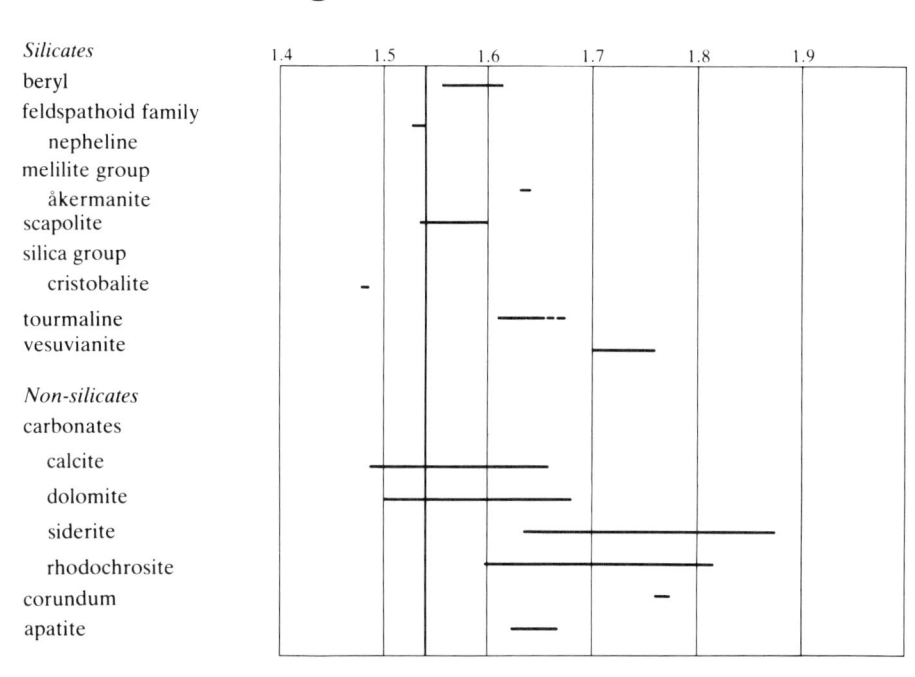

Silicates
beryl
feldspathoid family
  nepheline
melilite group
  åkermanite
scapolite
silica group
  cristobalite
tourmaline
vesuvianite

Non-silicates
carbonates
  calcite
  dolomite
  siderite
  rhodochrosite
corundum
apatite

Appendix A.4 Refractive indices of isotropic minerals

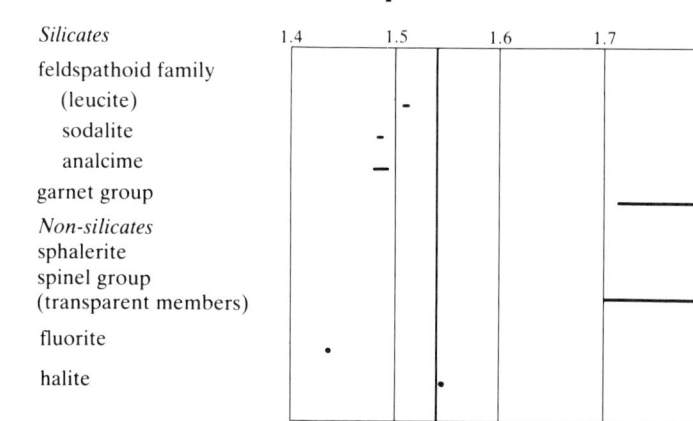

Silicates
feldspathoid family
  (leucite)
  sodalite
  analcime
garnet group

Non-silicates
sphalerite
spinel group
  (transparent members)
fluorite
halite

Appendix B 2V size and sign of biaxial minerals

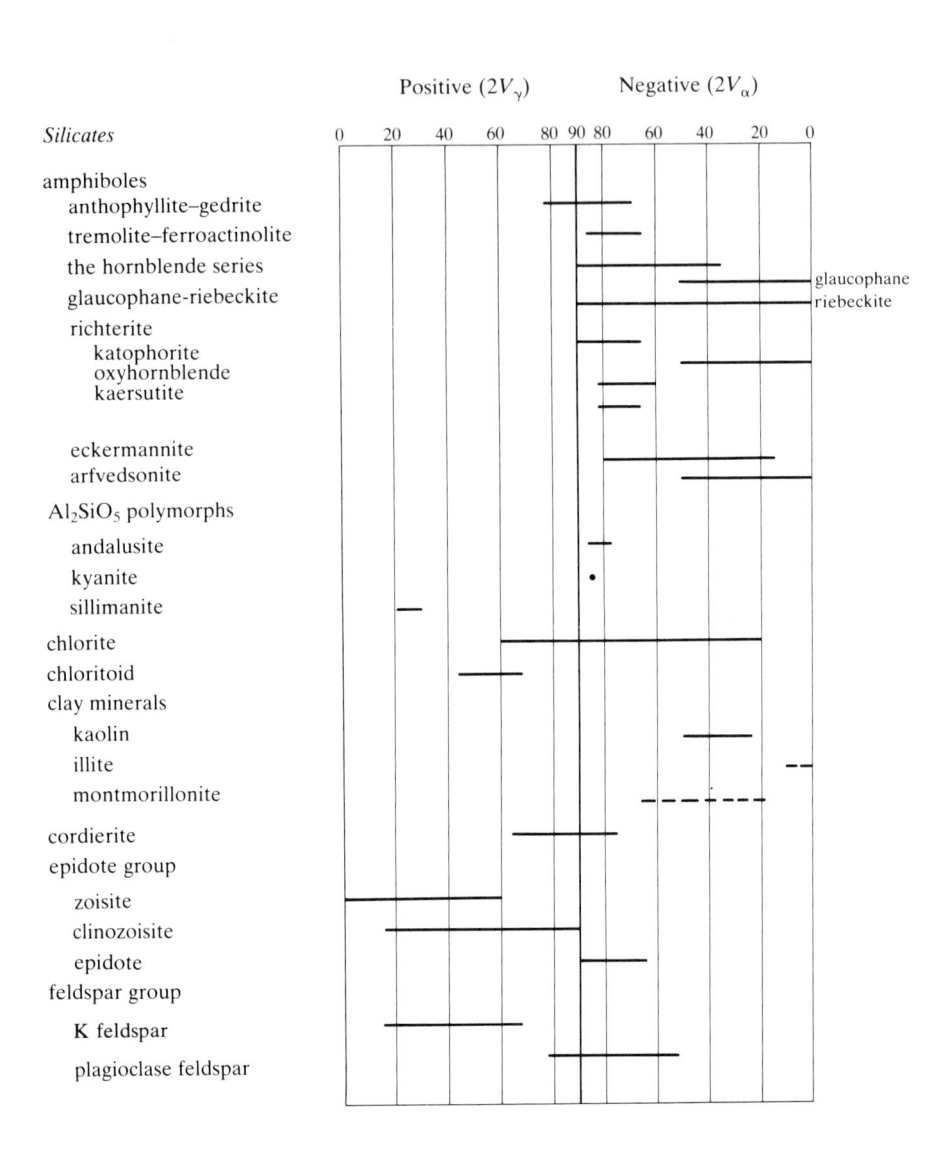

Positive ($2V_\gamma$)   Negative ($2V_\alpha$)

0 20 40 60 80 90 80 60 40 20 0

Silicates

amphiboles
  anthophyllite–gedrite
  tremolite–ferroactinolite
  the hornblende series
  glaucophane–riebeckite   glaucophane / riebeckite
  richterite
    katophorite
    oxyhornblende
    kaersutite
  eckermannite
  arfvedsonite

Al_2SiO_5 polymorphs
  andalusite
  kyanite
  sillimanite

chlorite
chloritoid
clay minerals
  kaolin
  illite
  montmorillonite

cordierite
epidote group
  zoisite
  clinozoisite
  epidote
feldspar group
  K feldspar
  plagioclase feldspar

	$2V_\gamma$						$2V_\alpha$			
	0	20	40	60	80	90 80	60	40	20	0

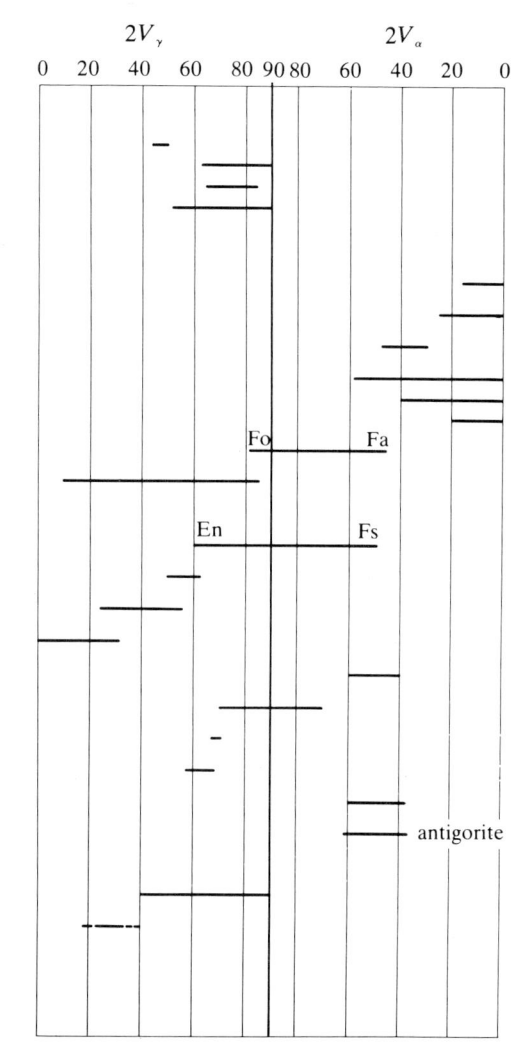

humite group
  norbergite
  chondrodite
  humite
  clinohumite

mica group
  phlogopite
  biotite
  muscovite
  lepidolite
  paragonite
  glauconite

olivine group — Fo ... Fa

pumpellyite

pyroxene group
  enstatite–orthoferrosilite — En ... Fs
  diopside–hedenbergite
  augite
  pigeonite
  aegirine
  aegirine–augite
  jadeite
  spodumene
  wollastonite

serpentine — antigorite

silica group
  tridymite

sphene

	$2V_\gamma$					$2V_\alpha$			
	0	20	40	60	80 90 80	60	40	20	0

staurolite
talc
topaz
zeolites

Non-silicates

carbonates
  aragonite
sulphates
  barite
  celestite
  gypsum
  anhydrite

wolframite

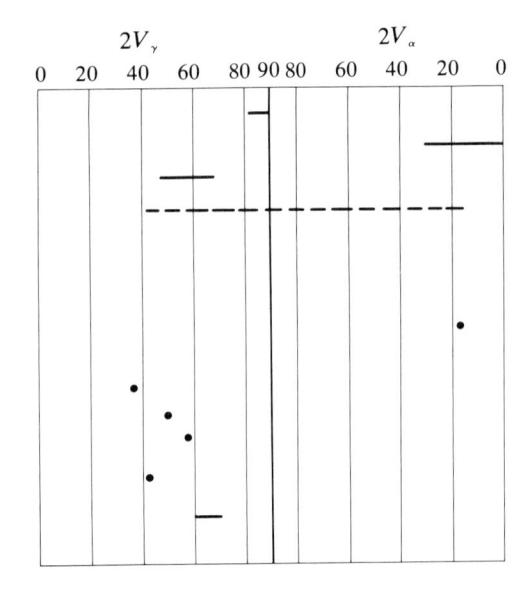

Appendix C Properties of ore minerals

Mineral	Formula	~R% range	Colour	~VHN	Anisotropy	Distinguished properties/resemblance (associations)	See page
acanthite	Ag_2S low temp. polymorph	30–31	light grey (greenish)	20–60	distinct	twin lamellae (Pb–Sb–As–Bi–Ag–Au)	—
alabandite	MnS	25	light grey	140–270	isotropic	lamellar twinning; brown or green internal reflections (sulphides, Mn-carbonates)	—
anatase	TiO_2 low temp. polymorph	20	light grey	580–620	weak	abundant colourful internal reflections; resembles rutile (Fe–Ti–O, pyrite)	164
argentite	Ag_2S high temp. polymorph	30	light grey (greenish)	20–60	isotropic	(Ag-sulphides, Cu–Pb–S, Au)	—
armalcolite	$(Fe,Mg)Ti_2O_5$	13–14	grey (brownish)	?	moderate	barrel shape (Fe–Ti–O, moon)	—
arsenopyrite	FeAsS	52	white	1050–1130	distinct	rhomb shape, twinning, zoning (Sulphides, oxides, Au–Bi–Te–Sn, W)	140
bismuth	Bi	60–65	bright white	10–20	distinct	tarnishes brown, multiple twinning (Co–Ni–As–S, Au–Bi, Te)	—
bismuthinite	Bi_2S_3	40–50	white (bluish)	70–220	very strong	fibrous, straight extinction (Co–Ni–As–S, Au–Bi–Te, Mo, Sn, W)	—

Mineral	Formula	~R% range	Colour	~VHN	Anisotropy	Distinguished properties/resemblance (associations)	See page
blaubleibender covellite	$Cu_{1+x}S$	↑	↑	↑	↑	resembles covellite but R_o blue in oil (Cu–Fe–S)	144
bornite	Cu_5FeS_4	22	light pinkish brown	100	very weak	tarnishes to blue or purple. Intergrowths with chalcopyrite (Cu–Fe–S)	141
boulangerite	$Pb_5Sb_4S_{11}$	38–41	light grey (greenish)	90–180	distinct	stronger anisotropy than bournonite (Pb–Sb–S, Cu–Fe–S)	—
bournonite	$CuPbSbS_3$	35–37	light grey (bluish)	130–210	weak	common twinning (Pb–Sb–S, Cu–Fe–S)	—
braunite	Mn_7SiO_{12}	20–22	light grey (brownish)	880–1190	weak	resembles some manganese oxides; may be magnetic (Mn–Fe–Si–O, mainly in metamorphic rocks)	—
bravoite	$(Fe,Ni,Co)S_2$	31→54 Ni,Co→Fe	light grey to white (brownish)	670–1540	isotropic	colour zoning (pyrite and other sulphides)	149
carrollite	Co_2CuS_4	43	white (pinkish)	350–570	isotropic	resembles linnaeite (Co–Cu–Fe–S)	—
cassiterite	SnO_2	11–13	grey	1030–1080	distinct	common twinning; strong colourless to brown internal reflections; cleavage (W, Bi, As, B, sulphides)	158
chalcocite	Cu_2S	32	light grey (bluish)	70–100	weak (colourful)	lancet shaped twinning (Cu–Fe–S)	142

mineral	formula	reflectance	colour	VHN	anisotropy	notes	page
chalcopyrite	CuFeS$_2$	42–46	yellow	190–220	weak	twinning; more yellow and softer than pyrite; as inclusions in sphalerite (Cu–Fe–Ni–S, sulphides)	143
chromite	FeCr$_2$O$_4$	12	grey (brownish)	1200–1210	isotropic (weak anisotropy)	rounded octahedra; resembles magnetite but non-magnetic (Fe–Ti–O)	160
cinnabar	HgS	28–29	light grey (bluish)	50–100	moderate	multiple twinning; abundant red internal reflections; rare (Hg–Sb–S, Fe–S)	144
cobaltite	CoAsS orthorhombic	53	white (pinkish)	1180–1230	weak	often idiomorphic cubic; colour zonation; cleavage traces (Cu–Fe–S, Co–Ni–As–S)	144
cohenite	Fe$_3$C		white		weak	resembles iron (Fe, Fe–O, Fe–Ni–S) meteorites	—
copper	Cu	81	metallic pink (tarnishes)	120–140	isotropic	scratches easily (Cu–O, Cu–Fe–S)	177
covellite	CuS	7–22	blue to bluish light grey	70–80	very strong (fiery orange)	plates and flakes; pleochroic (Cu–Fe–S)	144
cryptomelane	~K$_2$Mn$_8$O$_{16}$	27	light grey	530–1050	distinct	fibrous, botryoidal; resembles psilomelane; straight extinction (Fe–Mn–O)	—
cubanite	CuFe$_2$S$_3$ orthorhombic	40	light grey (yellowish brown)	150–260	strong	lamellae within pyrrhotite, chalcopyrite (Cu–Fe–S)	—

Mineral	Formula	~R% range	Colour	~VHN	Anisotropy	Distinguished properties/resemblance (associations)	See page
cubanite	CuFe$_2$S$_3$ cubic	35	light grey (pinkish)		isotropic	intergrown with orthorhombic cubanite (Cu-Fe-S)	—
cuprite	Cu$_2$O	25–30	light grey (bluish)	180–220	strong	deep red internal reflections (Cu, Fe-OH, Cu-Fe-S, Ag)	—
digenite	Cu$_9$S$_5$	22	light grey (bluish)	60–70	isotropic	(Cu-Fe-S)	142
djurleite	Cu$_{1.96}$S	↑	↑	↑	↑	resembles chalcocite	—
electrum	(Au,Ag)	83	light yellow	30–40	isotropic	resembles gold (Au-Te-Bi-Cu-Fe-As-Sb-Pb-S)	178
enargite	Cu$_3$AsS$_4$	25–29	light grey (pinkish)	130–580	strong colourful	cleavage ∥ (110) (Cu-Fe-Sb-As-S)	—
galena	PbS	43	white	70–80	isotropic	triangular cleavage pits (Pb-Ag-Sb-As-S, sulphides)	145
gersdorffite	(Ni,Co,Fe)AsS	47–54	white (pinkish)	520–910	isotropic	zoning, cleavage ∥ (100) gives triangular pits (Fe-Co-Ni-As-S)	—
glaucodot	(Co,Fe)AsS orthorhombic	45–50	white	840–1280	distinct	idiomorphic, cleavage; as inclusions in cobaltite (Co-Ni-As-S)	—
goethite	HFeO$_2$ orthorhombic	17	grey	770–820	distinct	Colloform, botryoidal or elongate crystals; red to brown internal reflections (in limonite, Fe-minerals, gossans)	170

mineral	formula	R%	color	VHN	anisotropy	properties	page
gold	Au	74	bright yellow	50	isotropic	very bright; as inclusions in sulphides; in fractures; soft (Au–Te–Bi–Cu–Fe–As–Sb–Pb–S)	178
goldfieldite	Cu$_3$(Te,Sb)S$_4$	32	light grey (brownish)		isotropic	zoned (Fe–Zn–S, Au–Ag–Te, Bi)	—
graphite	C	6–16	dark grey (brownish) to grey	10	strong	deformed flakes; bireflectance strong; cleavage (graphitic schists, graphite 'veins')	178
hematite	Fe$_2$O$_3$	25–30	light grey	920–1060	strong	tabular crystals, microcrystalline masses; lamellar twinning; weak bireflectance (Fe–Ti–O)	161
hydrocarbon		<5	dark grey		isotropic	rounded grains, interstitial masses, frosted surface, low reflectance but no internal reflections (sedimentary rocks, barite, carbonate, sulphide veins, U)	—
ilmenite	FeTiO$_3$	18–21	light grey (slightly pinkish)	520–700	moderate	occasional twinning; lamellar inclusions of hematite (Fe–Ti–O)	162
iron	Fe	65	bright white	120–290	isotropic	rounded grains (Fe–Ni–S, moon, meteorites)	—
jacobsite	(Mn,Fe,Mg)(Fe,Mn)$_2$O$_4$	19	grey (brownish-greenish)	720–750	isotropic	rounded grains; fine aggregates; strongly magnetic; resembles braunite (Mn-minerals; Fe–OH; in metamorphic rocks)	—

Mineral	Formula	~R% range	Colour	~VHN	Anisotropy	Distinguishing properties/resemblance (associations)	See page
jamesonite	$Pb_4FeSb_2S_8$	36–41	light grey (greenish)	70–130	strong	acicular with cleavage and twin lamellae parallel to length (Fe–Pb–Sb–Ag–S)	—
kamacite	(Fe,Ni)	60	white (bluish)		isotropic	(Fe–Ni–S, Fe–Ti–Cr–O, meteorites)	—
lepidocrocite	$FeO(OH)$	10–19	grey	690–780	very strong	red to brown internal reflections (In limonite, Fe-minerals, gossans)	170
limonite (see goethite and lepidocrocite)	$FeO.OH.nH_2O$	16–19	bluish grey	690–820	strong (colourful)	abundant brown to red internal reflections (replaces iron minerals)	170
linnaeite	Co_3S_4	45–50	white (pinkish)	350–570	isotropic	cleavage ∥ (100) (Cu–Fe–Ni–S)	—
livingstonite	$HgSb_4S_8$	35–40	light grey	70–130	strong (colourful)	scarce red internal reflections; more opaque than cinnabar (Hg–Sb–S)	—
loellingite	$FeAs_2$	55	white (yellowish)	370–1050	very strong (colourful)	common twinning (Fe–Ni–As–S, Cu–Fe–S, U, Sn)	—
mackinawite	$(Fe,Ni,Co,Cu)S$	22–45	light grey (pink red)	50–60	very strong	lamellae in Cu–Fe–S or Fe–Ni–S phases (Cu–Fe–Ni–S)	—
maghemite	$\gamma\text{-}Fe_2O_3$	26	light grey (bluish)	360–990	isotropic	magnetic (Fe–Ti–O)	—

Mineral	Formula	Reflectance	Colour	VHN	Anisotropy	Other properties	Page
magnetite	Fe_3O_4	21	light grey (often pinkish)	530–600	isotropic	magnetic; octahedra, rounded; lamellar inclusions of hematite or ilmenite (Fe–Ti–O)	163
manganite	MnOOH	15–21	grey (brownish)	370–800	strong	elongate crystal aggregates; cleavages; twin lamellae; red internal reflections; alters to pyrolusite (Fe–Mn–O–Si, veins)	—
marcasite	FeS_2 orthorhombic	49–55	white (slightly yellowish)	940–1290	strong (colourful)	radiating aggregates of twins; intergrown with pyrite (sulphides in sedimentary rocks, veins)	147
melnicovite–pyrite	$\sim FeS_2$	50	white (brownish)		isotropic or weak anisotropy	very fine grained aggregate; minute pits; banding (sulphide in sedimentary rocks)	150
miargyrite	$AgSbS_2$	30–50	light grey (bluish)	90–120	strong	twinning; scarce red internal reflections; as inclusions in galena (Ag–Pb–Sb–S)	—
millerite	NiS	52–56	yellow	190–380	strong	aggregates of acicular grains; common twinning (Cu–Fe–Ni–S)	—
molybdenite	MoS_2	15–37	grey to white	20–30	very strong	flakes, platelets, hexagonal; poor polish (Bi–Te–Au, Sn, W, sulphides)	147
niccolite	NiAs	52–58	white (orange or pinkish)	330–460	very strong	cataclased grains in radiating botryoidal masses; twinning (Ni–Co–Ag–As–U, sulphides)	176

Mineral	Formula	~R% range	Colour	~VHN	Anisotropy	Distinguishing properties/resemblance (associations)	See page
orpiment	As$_2$S$_3$	25	light grey	20–50	strong	radiating aggregates; abundant strong white to yellow internal reflections (realgar)	—
pentlandite	(Fe,Ni)$_9$S$_8$	44	white (slightly yellowish)	200–230	isotropic	triangular cleavage pits; alteration along octahedral parting; as lamellae in pyrrhotite (Cu–Ni–Fe–S)	148
perovskite	(Ca,Na,A)(Ti,Nb)O$_3$ A = rare earths	15	grey	920–1130	very weak	cubic octahedral habit; lamellar twinning; strong white to brown internal reflections (Cu–Fe–Ti–O, alkaline igneous rocks)	—
pitchblende	UO$_{2-3}$	16	grey	670–800	isotropic	radioactive; botryoidal masses, variation in R; shrinkage cracks (Ni–Co–Ag–Bi; Au, sulphides)	166
platinum	Pt	70	bright white (yellowish)	120–130	isotropic (incomplete extinction)	zoning; exsolved, intergrown phases of Pt group elements (Pt–Ir–Os–Rh–Ru–Pd, Fe–Cr–O, Cu–Fe–S)	—
proustite	Ag$_3$AsS$_3$	25–28	light grey (bluish)	110–140	strong	distinct bireflectance; red internal reflections (Ag–Pb–Sb–As–S)	148
pseudobrookite	Fe$_2$TiO$_5$	15	grey		distinct	red yellow internal reflections; resembles rutile (Fe–Ti–O)	—

mineral	formula	VHN	colour	reflectance	anisotropy	remarks	page
psilomelane	(Ba,Mn,Al,Fe,Si)$_3$Mn$_8$O$_{16}$·(O,OH)$_6$	15→30	grey→light grey (bluish)	200–810	strong	fine crystalline aggregates, botryoidal; resembles cryptomelane (Mn–O, Fe–OH; a weathering product)	—
pyrargyrite	Ag$_3$SbS$_3$	28–31	light grey (bluish)	50–130	strong	distinct bireflectance; red internal reflections (Ag–Pb–Sb–As–S)	148
pyrite	FeS$_2$ cubic	54	white (yellowish)	1030–1240	isotropic (weak anisotropy)	idiomorphic, fractured; framboids (common sulphide in all rock types)	149
pyrolusite	MnO$_2$	30–36	light grey (yellowish)	80–1500	very strong if well crystallised	coarse to cryptocrystalline; cleavage, twinning (Fe–Mn–O–OH; a weathering product; veins)	—
pyrrhotite	Fe$_{1-x}$S	40	white (brownish or pinkish)	230–320	strong (colourful)	polycrystalline aggregates, twinning; alters readily; magnetic (Cu–Fe–Ni–S, Fe–Ti–O)	150
rammelsbergite	(Ni,Co,Fe)As$_2$	60	bright white	460–830	strong (colourful)	radiating zoned aggregates, skeletal; common lamellar twinning (Co–Ni–Fe–As, Ag, Ag–Sb–S)	—
realgar	AsS	20–21	light grey	50–60	strong	abundant strong yellow to red internal reflections; interstitial (orpiment, Fe–As–Sb–S, Fe–OH)	—
rutile	TiO$_2$ high temp. polymorph	20–23	light grey (bluish)	1070–1210	strong	abundant bright internal reflections; twinning; acicular (Fe–Ti–O, Fe–S)	164

Mineral	Formula	~R% range	Colour	~VHN	Anisotropy	Distinguishing properties/resemblance (associations)	See page
safflorite	(Co,Fe,Ni)As$_2$	55–66	white (bluish)	430–990	strong (zoned)	concentric radiating aggregates with other minerals; star shaped twins (Co–Ni–Fe–As, Bi, U, Ag, sulphides)	—
scheelite	CaWO$_4$	10	dark grey	290–460	distinct	rhombohedral cleavages; abundant white internal reflections (wolframite, Cu–Fe–As–S, Bi–S, Sn, Au)	—
schreibersite	(Fe,Ni)$_3$P		white		weak	elongate inclusions in iron minerals (Fe–Ni–S, meteorites, moon)	—
silver	Ag	95	bright white (metallic)	50–120	isotropic	scratches easily; skeletal or dendritic (Co–Ni–Fe–As, oxidised sulphides)	179
skutterudite	(Co,Ni,Fe)As$_{3-x}$	53 variable	white (yellowish bluish or pinkish)	270–970	isotropic (weak anisotropy)	cleavage traces, compositional zoning (Co–Ni–F–As–S, Au, Ag, U, Mo, Bi)	—
sphalerite	(Zn,Fe)S	17	grey	190–210	isotropic	colourless to red internal reflections; irregular fractures and cleavage pits (sulphides)	151
spinel	MgAl$_2$O$_4$	8	dark grey	860–1650	isotropic	octahedral, rounded; internal reflections; as inclusions in magnetite (Fe–Ti–O)	165
stannite	Cu$_2$SnFeS$_4$	28–29	light grey (greenish to brownish)	140–330	strong	cleavage, triangular pits, pleochroism, lamellar and cross-hatch twinning (sulphides, Sn–W–As–Bi–Au)	—

Mineral	Formula						Page
stibnite	Sb_2S_3	30–40	light grey to white (slightly brownish)	40–110	very strong	acicular or bladed; distinct bireflectance; cleavage traces; deformation twinning (Ag–Pb–Sb–S, Fe–S, Au, Hg–S)	153
taenite	(Fe,Ni)		white (yellowish)		isotropic	lamellae in kamacite (Fe–Ni–S, Fe–Ti–Cr–O, meteorites)	—
tennantite	$Cu_{10}(Zn,Fe)_2(As,Sb)_4S_{13}$	31	light grey	320–370	isotropic	polycrystalline aggregates (Cu–Fe–S, galena, sphalerite)	154
tetradymite	Bi_2Te_2S	50–52	white (yellowish)	30–50	distinct	hexagonal cross sections, basal cleavage (Au–Bi–Te–S, Cu–Fe–S)	—
tetrahedrite	$(Cu,Ag)_{10}(Zn,Fe,Hg)_2(Sb,As)_4S_{13}$	31	light grey	320–370	isotropic	polycrystalline aggregates (Cu–Fe–S, galena, sphalerite)	154
titanomagnetite	$(Fe,Ti)_3O_4$	17	grey (brownish or pinkish)	720–730	isotropic (weak anisotropy)	homogeneous only if formed by rapid cooling, otherwise intergrown with Fe–Ti–O phases (Fe–Ti–O in igneous and metamorphic rocks)	163
todorokite	$(H_2O)_{<2}Mn_{<8}(O,OH)_{16}$	20	light grey	—	strong	columnar to fibrous aggregates, botryoidal; cleavage traces (Mn–Fe–O; deep sea Mn nodules)	—
troilite	FeS	17	yellow (brownish)		strong (colourful)	resembles pyrrhotite; rare (Fe–Ni–S, moon, meteorites)	—
ulvospinel	Fe_2TiO_4	17	brownish grey		isotropic	as fine intergrowths in titaniferous magnetite; octahedra (Fe–Ti–Cr–O, igneous, moon)	163

Mineral	Formula	~R% range	Colour	~VHN	Anisotropy	Distinguishing properties/resemblance (associations)	See page
uraninite	UO$_2$	17	grey	780–840	isotropic	associated with pitchblende (Ni–Co–Ag–Bi, Au)	166
wolframite	(Fe,Mn)WO$_4$	17	grey (slightly brownish)	360–390	moderate	bladed crystals, simple twins, reddish brown internal reflections (Sn, Au, Bi)	175
würtzite	ZnS hexagonal	↑	↑	↑		resembles sphalerite but rare	—

Appendix D Mineral identification chart

This simple chart shows the optical properties of the common ore minerals listed in order of relative polishing hardness, and it can be used as an aid to mineral identification. Reflectance values ($R\%$) are given with minerals plotted in their correct position, but some minerals are plotted in brackets in a second position because of their variable appearance.

Procedure

(1) Determine whether the unknown mineral is isotropic, weakly anisotropic or distinctly anisotropic. Weak anisotropy is seen using slightly uncrossed polars, whereas distinct anisotropy is easily visible with exactly crosssed polars.
(2) Note whether the mineral is colourless, slightly coloured or coloured in PPL; white to grey minerals are considered to be colourless for this purpose. Take care to consider the colour in relation to several adjoining minerals.
(3) Estimate the brightness (reflectance) of the mineral as a percentage. This is usually rather difficult unless some minerals in the section have already been identified. An uncertainty of ± 5 is typical for estimates of $R\%$ unless a good reference mineral is also in the section. If the mineral is distinctly bireflecting then estimate the minimum and maximum reflectance values.
(4) Whether the mineral is hard or soft can usually be determined by its polishing behaviour, e.g. pits persist in hard minerals, soft minerals scratch easily. Also, hard minerals tend to stand proud of the surface whereas soft minerals are scoured out. The Kalb light line can be used to compare polishing hardness with other minerals in the section.

If the properties of the unknown mineral appear to correspond to one of the minerals on the chart then it is best to check all the mineral's characteristics with the information given for the selected mineral in the description of the ore minerals before concluding the identification.

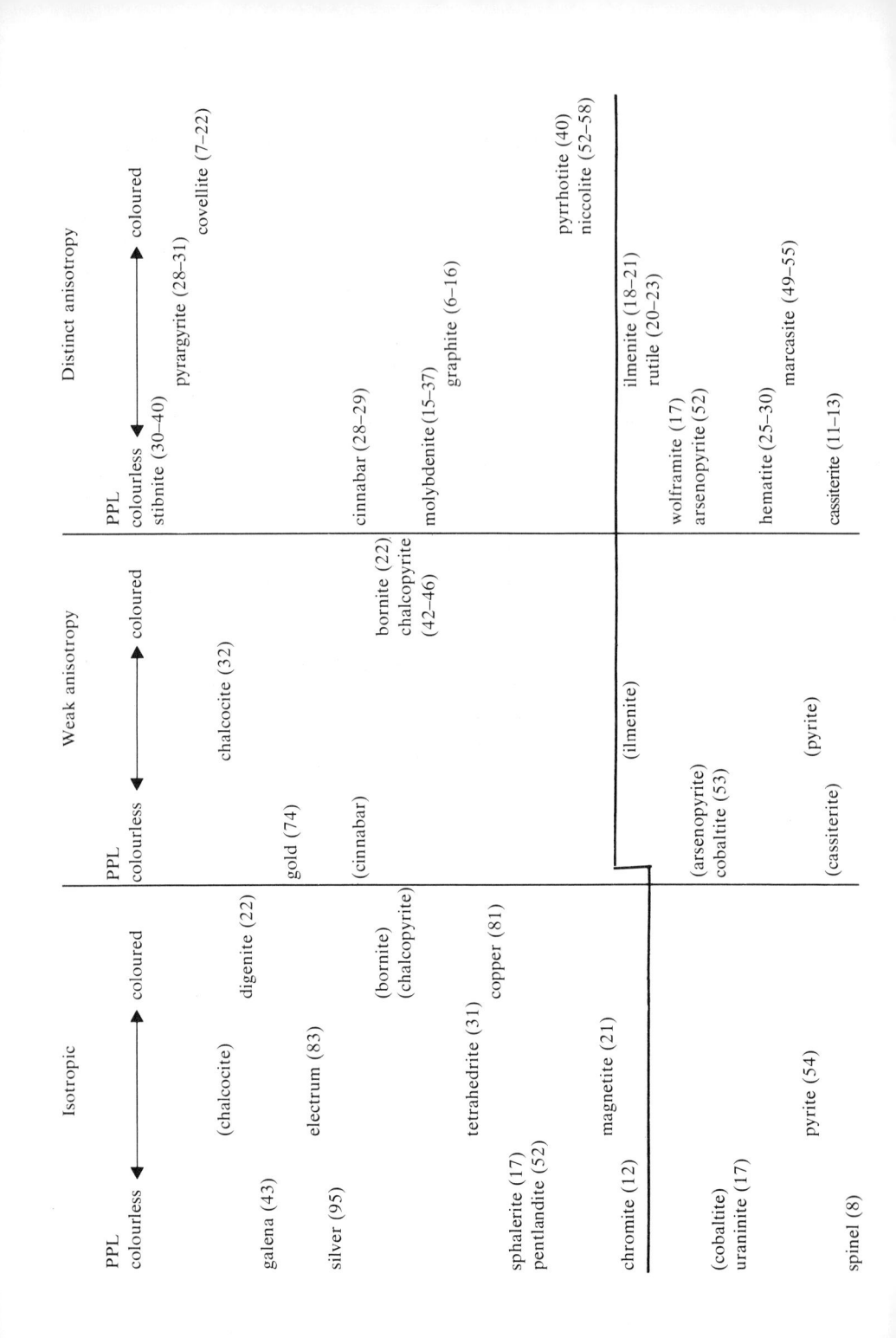

	Isotropic		Weak anisotropy		Distinct anisotropy	
	PPL colourless ↔	coloured	PPL colourless ↔	coloured	PPL colourless ↔	coloured
		(chalcocite)			stibnite (30–40)	pyrargyrite (28–31)
	galena (43)	digenite (22)				covellite (7–22)
	silver (95)	electrum (83)	gold (74)	chalcocite (32)		
			(cinnabar)		cinnabar (28–29)	
		(bornite) (chalcopyrite)		bornite (22) chalcopyrite (42–46)	molybdenite (15–37)	graphite (6–16)
	sphalerite (17) pentlandite (52)	tetrahedrite (31) copper (81)				pyrrhotite (40) niccolite (52–58)
	chromite (12)	magnetite (21)	(ilmenite)		ilmenite (18–21) rutile (20–23)	
	(cobaltite) uraninite (17)		(arsenopyrite) cobaltite (53)		wolframite (17) arsenopyrite (52)	
	pyrite (54)		(pyrite)		hematite (25–30) marcasite (49–55)	
	spinel (8)		(cassiterite)		cassiterite (11–13)	

Appendix E Gangue minerals

The gangue minerals referred to here are the minerals that commonly accompany ore minerals in hydrothermal deposits. Although they are transparent and are best studied using transmitted-light microscopy, it is useful to be able to *recognise* the common gangue minerals in reflected light (see Fig. 1.7). The minerals listed all have low reflectance values but the eye can determine small differences in brightness even at these low values. The carbonates are exceptional in having large birefringences and this results in distinct bireflectance; the resulting strong anisotropy is often masked by internal reflections. It is relatively easy to recognize a mineral as being a carbonate but it is difficult, as it is in thin section, to determine the type of carbonate. As well as using optical properties, physical and textural properties can be used in recognising the gangue minerals:

Quartz	Lack of cleavage but irregular fractures; crystal shape, especially pyramidal terminations; lack of alteration.
Barite	Several sets of cleavage traces; bladed or tabular crystals; radiating aggregates.
Fluorite	Octahedral cleavage giving up to three cleavage traces and triangular cleavage pits; cubic crystals.
K-feldspar	Several cleavage traces; alteration.
Carbonates	Rhombohedral cleavage resulting in up to three cleavage traces; multiple twinning; rhomb shaped crystals.

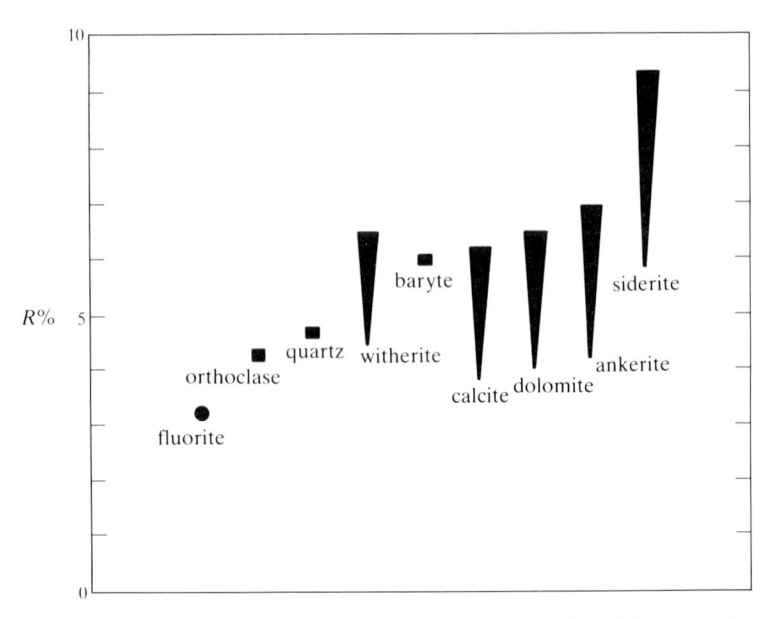

Note that carbonates have a pronounced bireflectance and R_o which is shown by *all* grains of a carbonate mineral is indicated by the wide end (maximum value) of the reflectance range.

Mineral	Refractive indices				Reflectances			
quartz, SiO_2	n_o	= 1.544	n_e	= 1.553	R_o	= 4.6	R_e	= 4.7
barite, $BaSO_4$	n_α	= 1.636	n_γ	= 1.648	R_α	= 5.8	R_γ	= 6.0
fluorite, CaF_2	n	= 1.434			R	= 3.2		
orthoclase, $KAlSi_3O_8$	n_α	= 1.518	n_γ	= 1.522	R_α	= 4.2	R_γ	= 4.3
calcite, $CaCO_3$	n	= 1.658	n_e	= 1.486	R_o	= 6.1	R_e	= 3.8
dolomite, $CaMg(CO_3)_2$	n_o	= 1.679	n_e	= 1.500	R_o	= 6.4	R_e	= 4.0
ankerite, $Ca(Fe,Mg)(CO_3)_2$	n_o	= 1.710	n_e	= 1.515	R_o	= 6.9	R_e	= 4.2
siderite, $FeCO_3$	n_o	= 1.875	n_e	= 1.635	R_o	= 9.3	R_e	= 5.8
witherite, $BaCO_3$	n_α	= 1.529	n_γ	= 1.677	R_α	= 4.4	R_γ	= 6.4

Bibliography

Atkin, B. P. and P. K. Harvey 1982. NISOMI-81: an automated system for opaque mineral identification in polished section. *Process mineralogy II: applications in metallurgy, ceramics and geology.* Conference Proceedings, The Metallurgical Society of AIME, Dallas, 77–91.

Bastin, E. S. 1950. *Interpretation of ore textures.* Mem. Geol. Soc. Am. **45**

Bloss, F. D. 1971. *Crystallography and crystal chemistry.* New York: Holt Rinehart and Winston.

Bowen, N. L. and J. F. Schairer 1935. The system $MgO-FeO-SiO_2$. *Am. J. Sci.* **229**, 151–217.

Bowie, S. H. U. and P. R. Simpson 1977. Microscopy: reflected-light. In *Physical methods in determinative mineralogy*, 2nd edn, J. Zussman (ed.). London: Academic Press.

Bowie, S. H. U. and K. Taylor 1958. A system of ore mineral identification. *Mining Mag. (Lond.)* **99**, 265–77, 337–45.

Buddington, A. F. and D. H. Lindsley 1964. Iron-titanium oxide minerals and synthetic equivalents. *J. Petrol.* **5**, 310–57.

Cameron, E. N. 1961. *Ore microscopy.* New York: John Wiley.

Cervelle, B. 1979. The reflectance of absorbing anisotropic minerals. In *Proceedings of the 1974 ore microscopy summer school at Athlone*, M. J. Oppenheim (ed.). Special Paper no. 3, 49–57. Geol. Surv. of Ireland.

Commission of Ore Microscopy 1977. *IMA/COM quantitative data file*, N. F. M. Henry (ed.). London: Applied Mineralogy Group, Mineralogical Society.

Craig, J. R. and D. J. Vaughan 1981. *Ore microscopy and ore petrography.* New York: Wiley.

Deer, W. A., R. A. Howie and J. Zussman 1962. *Rock forming minerals*, Vols 1–5. London: Longmans.

Deer, W. A., R. A. Howie and J. Zussman 1966. *An introduction to the rock-forming minerals.* London: Longmans.

Deer, W. A., R. A. Howie and J. Zussman 1978. *Rock-forming minerals: single chain silicates*, Vol. 2A. London: Longmans.

Edwards, A. B. 1947. *Textures of the ore minerals.* Melbourne: Austral. Inst. Min. Metall.

Freund, H. (ed.) 1966. *Applied ore microscopy.* New York: Macmillan.

Galopin, R. and N. F. M. Henry 1972. *Microscopic study of opaque minerals.* London: McCrone Research Associates. Originally published by Heffer, Cambridge.

Hallimond, A. F. 1970. *The polarising microscope.* York: Vickers Instruments.

Henry, N. F. M. (ed.) 1970–8. *Mineralogy and materials news bulletin for microscopic methods.* London: Applied Mineralogy Group, Mineralogical Society.

Holdaway, M. H. 1971. Stability of andalusite and the aluminium silicate phase diagram. *Am. J. Sci.* **271**, 97–131.

IMA/COM 1977. *Quantitative data file, First Issue. International Mineralogical Association, Commission on Ore Microscopy.* London: McCrone Research Associates.

Judd, D. B. 1952. *Colour in business, sciences and industry.* New York: John Wiley.

Kerr, P. F. 1977. *Optical mineralogy*. New York: McGraw–Hill.
Leake, B. E. 1978. Nomenclature of amphiboles. *Can. Mineral.* **16**, 501–20.
Lister, B. 1978. *Ore polishing*. Institute of Geological Sciences Report 78/27. London: HMSO.
Nickless, G. (ed.) 1968. *Inorganic sulfur chemistry*. Amsterdam: Elsevier.
Oppenheim, M. J. (ed.) 1979. *Proceedings of the 1974 ore microscopy summer school at Athlone*. Geol. Surv. of Ireland, Special Paper no. 3.
Palache, C., H. Berman and C. Frondel (eds) 1962. *Dana's system of mineralogy*, 7th edn. Vols I, II, III. New York: John Wiley.
Pauling, L. and E. H. Neuman 1934. The crystal structure of binnite, $(Cu,Fe)_{12}As_4S_{13}$ and the chemical composition and structure of minerals of the tetrahedrite group. *Z. Krist.* **88**, 54–62.
Phillips, W. R. and D. T. Griffin 1981. *Optical mineralogy. The non-opaque minerals*. San Francisco: W. H. Freeman.
Picot, P. and Z. Johan 1977. *Atlas des minéraoux metalliques*. Mém. B.R.G.M., Orléans, no. 90.
Piller, H. 1966. Colour measurements in ore microscopy. *Mineral. Deposita.* **1**, 175–92.
Piller, H. 1977. *Microscope photometry*. Berlin: Springer–Verlag.
Ramdohr, P. 1969. *The ore minerals and their intergrowths*. Oxford: Pergamon.
Ribbe, P. H. 1974. *Short course notes*, Vol. 1: *Sulfide mineralogy*. Mineralogical Society of America.
Rumble, D. III (ed.) 1976. *Short course notes*, Vol. 3: *Oxide minerals*. Mineralogical Society of America.
Shuey, R. T. 1975. *Semiconducting ore minerals. Developments in economic geology*, no. 4. Amsterdam: Elsevier.
Smith, J. V. 1974. *Feldspar minerals*, Vols 1–3. Berlin: Springer–Verlag.
Tuttle, O. F. and N. L. Bowen 1958. Origin of granite in the light of experimental studies in the system $NaAlSi_3O_8$–$KAlSi_3O_8$–SiO_2–H_2O. *Mem. Geol. Soc. Am.* **74**.
Uytenbogaardt, W. and E. A. J. Burke 1971. *Tables for microscopic identification of ore minerals*. Amsterdam: Elsevier.
Vaughan, D. J. and J. R. Craig 1978. *Mineral chemistry of metal sulfides*. Cambridge: Cambridge University Press.
Wahlstrom, E. E. 1959. *Optical mineralogy*, 2nd edn. New York: John Wiley.

Index

The pages on which minerals and other terms are most directly introduced or defined are shown in **bold** type. Numbers in *italics* refer to text illustrations. Reference is also made to tables, appendices and main text sections.

abnormal interference colour **192**
absorption coefficient **204**
acanthite App. C
accessory slot **4**
actinolite (ferroactinolite) **46–7**
aegirine, aegirine–augite 11, **114**
aegnigmatite **56**
åkermanite **90**
Al$_2$SiO$_5$ polymorphs 30, **35–40**
 andalusite 35
 kyanite 37
 sillimanite 39–41
alabandite App. C
albite twins 75, **80**
alkali feldspars 34, **72–7**
 microcline–low albite 73–4
 orthoclase–low albite 73–4
 sanidine–high albite 72
allanite 63
almandine **87–8**
alteration 7
amosite 45
amphibole group 11, 31, **41–56**
 alkali amphiboles 51–6
 Ca-poor amphiboles 43–5
 Ca-rich amphiboles 46–50
amphibolisation 47
analcime **86–7**
analyser **4**, 16
anatase 164, 165, App. C
andalusite 6, 10, 30, **35**
andradite **87–8**
anhydrite **171**, 175
anisotropic
 crystals (transmitted light) 8, 183
 sections (reflected light) 212
anisotropy
 reflected light 20, **212–17**
ankerite 137, 240
annabergite 177
annite 93
anomalous interference colour **192**
anthophyllite–gedrite **43**
antigorite 119
antiperthite **70**, 82
apatite **175**, Table 1.2
aperture diaphragm 3, 16, *1.6*

aquamarine 56
aragonite **134**, Table 3.1
arfvedsonite **55**, Table 4.10
argentite App. C
armalcolite App. C
arsenide **176**
 niccolite 177, App. C
arsenopyrite **139**, App. C
asbestos minerals
 amosite 45
 anthophyllite 43
 chrysotile 119
 crocidolite 52
augite 11, **110–11**

barite **172**, 240
barkevikite 55
Barrovian-type metamorphism 40
baryte(s) *see* barite
Bertin's surface 194
Bertrand lens **4**, 16, 32
beryl 30, **56**
biaxial figures 9, **192–7**
biaxial indicatrix **183**
biotite 5, **92–3**
 annite 93
 lepidomelane 93
 siderophyllite 93
bireflectance **20**
birefringence **8–9**
bismuth App. C
bismuthinite App. C
blaubleibender covellite 145, App. C
Blue John 169
blueschists 52
bornite **141**, App. C
boulangerite App. C
bournonite App. C
braunite App. C
bravoite **150**, App. C
brookite 164, 165
brucite 137, 169
'Buchan'-type metamorphism 40, 62
buffing 29

calcite **136**, 240, *3.1*, Table 1.2, Table 3.1
cancrinite 84

INDEX

carbonates **132–8**, *3.2*
   ankerite 137
   aragonite 136, Table 3.1
   calcite 134, Table 3.1
   dolomite 137, Table 3.1
   rhodochrosite Table 3.1
   siderite 137, Table 3.1
   strontianite Table 3.1
   witherite Table 3.1
carlsbad twins 80
   carlsbad–albite twinning 80
carrollite App. C
cassiterite **158–9**, App. C
celestine *see* celestite
celestite **172**
celsian feldspar **84**
chain silicates 30–1
   single 31, *2.1*
   double 31, *2.2*
chalcocite **142**, App. C
chalcopyrite **143**, App. C
chalcopyrrhotite 151
chlorite 34, **57–8**
chloritoid 30, **58**
chondrodite 88
chromite **160**, App. C, Table 3.2
chrysotile 119
cinnabar **144**, App. C
clay minerals 8, 32, **59–60**
   illite 60
   kaolin 59
   montmorillonite 60
   smectite 60
   vermiculite 60
cleavage
   reflected light 21
   transmitted light **6**
clinohumite **88–9**
clinopyroxenes 11, 99–102, 103, **108–17**
clinozoisite **64–5**
cobaltite **144**, App. C
cohenite App. C
collophane 175
colour (of a mineral)
   reflected light 19, **209–11**
   transmitted light 5
colour diagram (CIE 1931) **210–11**
components (of light) **190**
condenser **4**
convergent lens **3**
copper **177**, App. C
cordierite 30, **61–2**
corundum **160**, Table 1.2
covelline *see* covellite
covellite **144–5**, 210, App. C
cristobalite **123**

crocidolite 52
cross-hatched twinning 75
crossed polars
   adjustment 27, 28
   reflected light 20, 213
   transmitted light 4
crossite **51**, Table 4.10
cryptomelane App. C
cryptoperthite 70, 73
crystal symmetry **206**, *5.4*
cubanite App. C
cummingtonite–grunerite **45**
cuprite 178, App. C
cyclosilicates 30
   ring silicates 30

desert rose 175
determination of order of colour 191
diamond Table 1.2
diaphragm **3**
digenite **142**, App. C
diopside–hedenbergite 11, **108**
dispersion **12**
dispersion curves 204
djurleite 143, App. C
dolomite **137**, Table 3.1, 240
dominant wavelength 210

eckermannite–arfvedsonite **55–6**
eclogite 88
edenite 47
electrum 178, App. C
emerald 56
enargite App. C
enstatite–orthoferrosilite **105–7**
epidote group 30, **63–7**
   clinozoisite 64–5
   epidote (pistacite) 66–7
   zoisite 63–4
exsolution lamellae **102–3**
extinction angle 10–11, **200–1**
eyepiece **4**

fayalite **95–7**
feldspar group 7, 34, **67–84**
   alkali feldspars 72–7
   celsian 84
   plagioclase feldspars 78–84
feldspathoid family 34, **84–7**
   analcime 86
   cancrinite 84
   hauyne 84
   kalsitite 84
   leucite 84
   nepheline 85–6
   nosean 84
   sodalite 86

INDEX

ferberite 175
ferrianilmenite 156
ferritchromite 160
ferropseudobrookite 156
field diaphragm 3, 17
first-order red plate **191**
flash figures **196–7**
fluorite **168**, 240, Table 1.1, Table 1.2
focusing (of microscope) **5**
forsterite **95–7**
framboidal pyrite 149
framework silicates 34
freibergite 154
Fresnel equation **203–6**

galena **145**, App. C, Table 1.1
gangue minerals App. E
garnet group 6, 30, **87–8**
gedrite **43**
gehlenite **90**
geikielite 162
gersdorffite App. C
glaucodot App. C
glaucophane–riebeckite **51–2**
goethite **170**, App. C
gold **178**, App. C
goldfieldite App. C
graphite **178**, App. C
grossular **87–8**
grunerite **45**
gypsum 171, **174**, Table 1.2

habit **6**
halides **167–9**
  fluorite 168
  halite 169
halite **169**
  structure *3.15*
hastingsite 47
hauyne 84
hematite **161**, App. C, Table 1.1
hemo-ilmenite 156
hercynite 165–6, Table 3.2
high albite **72–7, 78–81**
hornblende series 11, **47–50**
  edenite 47
  hastingsite 47
  pargasite 47
  tschermakite 47
hue (of colour) 210
huebnerite 175
humite **88–9**
humite group 30, **88–9**
hydrocarbon App. C
hydrogrossular **87**
hydromagnesite 169

hydroxides **169–71**
  brucite 169
  goethite 170
  lepidocrocite 170
  limonite 170

idocrase 30, **129**
illite 32, **60**
ilmenite 156, **162**, App. C
ilmeno-hematite 156
incident illuminators **13–14**, *1.4*
inclusions 21
inosilicates 30
  chain silicates 30
  single chain silicates 31, *2.1*
  double chain silicates 31, *2.2*
interference colours 8–9, **186–9**
interference figure 9–10, **192–6**, *4.20, 4.21*
internal reflections **21**
iron App. C
iron–titanium oxides **155–6**
isochromatic curves 194
isogyres **194–5**
isotropism **8**
isotropic
  crystals (transmitted light) 181
  sections (reflected light) 213

jacobsite App. C
jadeite **112**
jamesonite App. C

K-feldspar **72–7**
kaersutite **54**
Kalb light line **24**, *1.7*
kalsilite 84
kamacite App. C
kaolin(ite) 32, **59**
katophorite **54**, Table 4.10
kyanite 30, **37**, Table 4.10

lepidocrocite **170**, App. C
lepidomelane 93
leucite **84–5**
leucoxene 124, 162
light source **1**
limonite **170**, App. C
linnaeite App. C
livingstonite App. C
lizardite 119
loellingite App. C
low albite **73–7, 78–81**

mackinawite App. C
maghemite 157, App. C, Table 3.2
magnetite 156, **163**, App. C, Table 3.2

INDEX

manganite App. C
marcasite **147**, App. C
martite 162
melilite group 30, **90**
melnikovite 150, App. C
mesoperthite 70
metacinnabarite 144
miargyrite App. C
mica group 34, **90–4**
  biotite 92–3
  glauconite 91
  lepidolite 91
  muscovite 94
  paragonite 91
  phlogopite 91–2
microcline **74–7**
microhardness **25**
micro-identation hardness *1.9*
microscope, reflected light **12–16**
  analyser 16
  Bertrand lens 16
  incident illuminators 13–14
  light control 16–17
  light source 12
  objectives 15–16
  polariser 12
  use of 26–8
microscope, transmitted light **1–5**
  accessory slot 4
  analyser 4
  Bertrand lens 4
  condenser 3
  convergent lens 3
  crossed polars 4
  eyepiece 4
  focusing 5
  light source 1
  objectives 4
  plane polarised light 3
  polariser 1
  stage 4
  sub-stage diaphragms 3
  use of 27–8
migmatite 40
millerite App. C
mineral properties, transmitted light **5–12**
  alteration 7
  birefringence 8–9
  cleavage 6
  colour 5
  dispersion 12
  extinction angle 10–11
    maximum extinction angle 11
    oblique 10
    straight 10
  habit 6

interference colour 8–9
interference figures 9–10
  biaxial 9–10
  isotropism 8
  pleochroism 5
  relief 6
  twinning 11
  uniaxial 9
  zoning 11–12
mispickel *see* arsenopyrite
Moh's hardness Table 1.2
molybdenite **147**, App. C
monochromatic (light) 186, *4.1*
montmorillonite group (smectites) 32, **60**
muscovite **94**
  sericite 94

native elements **177–9**
  copper 177
  gold 178
  graphite 178
  silver 179
natrolite **130**
nepheline **85–6**
nephrite 47
nesosilicates 30
  island silicates 30
  orthosilicates 30
Newton's scale 186, **189**
niccolite **176–7**, App. C
nickeline *see* niccolite
norbergite **88–9**
nosean 84
numerical aperture 15, *1.5*

objectives (lenses) **4**, **15–16**
  cleaning 23
oblique extinction **201**
oil immersion
  reflected light **22–3**
olivine group 8, 10, 30, **95–7**
  corona structures 97
  kelyphitic margin 97
omphacite **113**
ophiolite suites 52
optic axes **184**
optic axial angle 10, **184**
optic axial plane **184**
optical constants **203**
  dispersion curves of 204
orpiment App. C
orthoclase 67–70, **73–7**, 240, Table 1.2
orthopyroxenes 10, 99, 102–4, **105–7**
oxides **155–67**
  cassiterite 158–9
  chromite 160

INDEX

oxides *cont.*
  corundum 160
  hematite 161
  ilmenite 162
  iron–titanium oxides 155–6
  magnetite 163
  rutile 164–5
  spinel group 157–8
  spinel 165
  uraninite 166–7
oxyhornblende **54**

pargasite 47
path difference 186
pentlandite **148**, App. C
periclase 137, 169
perthite **70**, 77
perovskite App. C
phlogopite **91–2**, 93
phosphate **175**
  apatite 175
  collophane 175
phyllosilicates 32
  polytypes 32
  sheet silicates 32, *2.3*
picotite Table 3.2
piemontite 63
pigeonite **109**, Table 4.10
pitchblende 166, App. C
plagioclase feldspars 11, **78–84, 90–3**
  albite 78
  albite twinning 80
  anorthite 78
  carlsbad twinning 80
  celsian 84
  lamellar twinning 80
plane polarised light **3**
platinum App. C
pleochroic schemes 5, **197–201**
pleochroism
  reflected light **19**
  transmitted light **5–6**
pleonaste Table 3.2
polarisation
  colours **212–17**
  figures 16
polarised light **180**
  circularly 202, *5.1*
  elliptically 202, *5.1*
  linearly 180, 202, *5.1*
  plane 180, *5.1*
polariser **1**
polished sections
  appearance 17
  preparation 28, *1.11*
  systematic description 19

polishing
  hardness **23**
  procedure 28
  relief 18, **23**
potassium feldspar **72–7**
polytypes 32
prehnite 34
proustite **148**, App. C
pseudobrookite 156, App. C
pseudoleucite 85
psilomelane App. C
pumpellyite 30, **98**
pyrargyrite **148**, App. C
pyrite **149–50**, App. C
pyrolusite App. C
pyrope **87–8**
pyrophanite 162
pyroxene group 6, 31, **99–117**
  aegirine (acmite) 114
  aegirine–augite 114
  augite 110–11
  crystallisation trends 103–4
  diopside–hedenbergite 108–9
  enstatite–orthoferrosilite 105–7
  exsolution lamellae 102
  jadeite 112
  monoclinic pyroxenes (cpx) 99, 108–20
  omphacite 113
  orthopyroxenes (opx) 99, 105–7
  pigeonite 109
  spodumene 116
  wollastonite 116–17
pyrrhotine *see* pyrrhotite
pyrrhotite **150–1**, App. C

quantitative colour **209–11**
quartz 6, 34, **121–2**, 240, Table 1.2
  α-quartz 120–1
  β-quartz 120–1
quartz wedge **191**

rammelsbergite App. C
realgar App. C
reflectance 19, **203–9**
  grey scale **19**
  indicating surfaces 206–9
reflected light
  microscope **12–16**
  theory **202–17**
refractive index 6, **181**, 203
relative polishing hardness **23–5**, *1.8*
relief
  reflected light **24**
  transmitted light **6**
resolution *1.5*
retardation **187**

INDEX

rhodochrosite Table 3.1
richterite **53**
riebeckite **51–2**
rotation (of polarised light) **213–17**
   colours **213–17**
   exercise *5.8*
ruby 160
ruby silvers 148–9
rutile 156, **164–5**, App. C

safflorite App. C
sanidine **72–7**
sapphire 160
saturation (colour) 210
saussuritisation 80
scapolite 34, 117–18
scheelite 176, App. C
schriebersite App. C
scratch hardness (Moh's) 26
sensitive tint **191**
sericite 94
serpentine 7, 34, **119**
   antigorite 119
   chrysotile 119
   lizardite 119
sheet silicates 32, *2.3*
siderite **137–8**, 240, Table 3.1
siderophyllite 93
sign determination
   biaxial **196**, *4.20*
   uniaxial **197**, *4.21*
silica group **120–3**
   coesite 121
   cristobalite 123
   quartz 121–2
   stishovite 121
   tridymite 122
sillimanite 10, 20, **39–40**
silver **179**, App. C, Table 1.1
skutterudite App. C
sodalite **86**
sorosilicates 30
spectral reflectance curves 204
spessartine **87–8**
sphalerite **151–3**, App. C, Table 1.1
sphene 30, **124**
   leucoxene 124
spinel **165–6**, App. C, Table 3.2
spinel group **157–8**, Table 3.2
   solid solution *3.10*
   unit cell *3.9*
spodumene **116**
stage (microscope) **4**
standard wavelengths 12, 209
stannite App. C
staurolite 30, **125–6**

stibnite **153**, App. C
straight extinction 10, **201**
strontianite Table 3.1
substitution, coupled 34
sulphates **172–5**
   anhydrite 171–2
   barite 172–3
   celestite 172–3
   gypsum 174–5
sulphides **138–54**
   arsenopyrite 139–41
   bornite 141–2
   chalcocite 142–3
   chalcopyrite 143
   cinnabar 144
   cobaltite 144
   covellite 144–5
   digenite 142
   galena 145–6
   marcasite 147
   molybdenite 147
   pentlandite 148
   proustite 148–9
   pyrargyrite 148–9
   pyrite 149–50
   pyrrhotite 150–1
   sphalerite 151–2
   stibnite 153
   structures *3.3*
   tennantite 154
   tetrahedrite 154
sulphosalts 138–9, 146

taenite App. C
talc 34, **126–7**, Table 1.2
tarnishing 18, 209
tektosilicates 34
   coupled substitution 34
   framework silicates 34
tennantite **154**, App. C
tetradymite **148**, App. C
tetrahedrite **154**, App. C
thin section
   preparation 28, *1.11*
thucolite 167
titanohematite 156
titanomaghemite 156
titanomagnetite 156, App. C
todorokite App. C
topaz 30, **127–8**, Table 1.2
tourmaline 30, **128–9**
   dravite 128
   elbaite 128
   schorl 128
tremolite–(ferro) actinolite 11, **46–7**
tridymite **122–3**

INDEX

troilite 150, App. C
tschermakite 47
tungstate **175–6**
  wolframite 175–6
tungstenite 148
twinning
  reflected light 21
  transmitted light 11

ulvospinel 163, App. C, Table 3.2
uniaxial figures 9, **197–200**
uniaxial indicatrix **184**
uralitisation 47
uraninite **166–7** App. C

vermiculite **60**
vesuvianite 30, **129**
  idocrase 129
Vickers hardness number (VHN) 22, **25–6**, Table 1.2

visual brightness **210**

witherite 240, Table 3.1
wolfram *see* wolframite
wolframite **175–6**, App. C
wollastonite 32, **117**
wood tin 159
wurtzite 153, App. C
wüstite 156

zeolite group 34, 84, **129–30**
  analcime–natrolite 130
  chabazite–thomsonite 130
  mesolite–scolecite 130
zircon 30, 131
zoisite **63–4**
zoning
  reflected light 21
  transmitted light **11**